Oxford Shakespeare Topics

GENERAL EDITORS: LENA COWEN ORLIN, PETER HOLLAND,
AND STANLEY WELLS

Shakespeare and Science

TOM RUTTER

OXFORD
UNIVERSITY PRESS

OXFORD
UNIVERSITY PRESS

Great Clarendon Street, Oxford, OX2 6DP,
United Kingdom

Oxford University Press is a department of the University of Oxford.
It furthers the University's objective of excellence in research, scholarship,
and education by publishing worldwide. Oxford is a registered trade mark of
Oxford University Press in the UK and in certain other countries

© Tom Rutter 2024

The moral rights of the author have been asserted

Published in the United States of America by Oxford University Press
198 Madison Avenue, New York, NY 10016, United States of America

British Library Cataloguing in Publication Data
Data available

Library of Congress Control Number: 2024937954

ISBN 9780192898548
ISBN 9780192898555(pbk.)

DOI: 10.1093/oso/9780192898548.001.0001

Printed and bound by
CPI Group (UK) Ltd, Croydon, CR0 4YY

Links to third party websites are provided by Oxford in good faith and
for information only. Oxford disclaims any responsibility for the materials
contained in any third party website referenced in this work.

Acknowledgements

I would like to thank Eleanor Collins at Oxford University Press for believing in this project from the start, and the Series Editors for supporting it. I am grateful to the Press's anonymous reader for providing helpful suggestions from the proposal stage onwards, to Alexander Hardie-Forsyth for helping the manuscript through review, and to Gangaa Radjacoumar and all others involved in the production process.

Much of this book was researched and written in the period 2020–21 when mobility, and so much else, was constrained by lockdown measures designed to combat the Covid-19 virus. I am grateful to all those who made work possible under these conditions, including the library staff at the University of Sheffield and the digital support staff (especially James Pearson) who helped Sheffield's early modern research culture to continue to thrive in the online sphere. Thanks to Ana D. Rodriguez of the Rare Book and Manuscript Library, University of Illinois at Urbana-Champaign for photographing Figure 4.1. The School of English at the University of Sheffield provided me with a period of research leave in spring 2021: thanks to Jane Hodson for this and for leading the School during a period of unprecedented stress and disruption. In the wake of lockdown it was revitalizing to be able to present research connected with this book in person. Thanks to Namratha Rao and Ted Tregear for convening the 'Material and Metaphysics' panel at the Renaissance Society of America conference in Dublin in 2022, and to Todd Borlik, Darryl Chalk, and Laurie Johnson for the 'Re-wilding Shakespeare' seminar at the British Shakespeare Association in Liverpool in 2023.

Over the last few years, professional and personal networks have been especially important. I am enormously grateful to my wonderful early modern colleagues in the School of English, Marcus Nevitt, Anna Reynolds, Emma Rhatigan, and Cathy Shrank, for advice and encouragement, and for invaluable suggestions at the Pre-modern Work in Progress session. Beyond English, my colleagues in the

Sheffield Centre for Early Modern Studies have helped to provide a vital interdisciplinary context for research into the early modern period: thanks in particular to Jamie Graves, Cora James, Tom Leng, Lucy Morgan, Emily Naish, Leonie Price, Mabel Winter, and Phil Withington. I have been inspired by the talent and dedication of the PhD students with whom I've worked while writing this book: Dorothy Bowles, Rama Friedlander, Amy Jackson, Cora James, Grace Mold, Emily Naish, and Melanie Russell. My co-editors at the journal *Shakespeare*, Deborah Cartmell, Brett Greatley-Hirsch, and Lisa Hopkins, have also been an invaluable source of support.

As always, this book would not have been written without the love, encouragement, and bracing mockery of Aphra, Caedmon, and Sophie.

My father, William Rutter—teacher, artist, fly-fisherman, cook, raconteur—died while my research on this project was in its later stages. Years ago, when I was completing my PhD thesis, he suggested with a characteristic blend of humour, self-importance, and cavalier disregard for factual accuracy that I should dedicate it 'To my Dad—he taught me all I know'. At the time, for reasons that can be imagined, this was not a request I felt able to honour. Two decades later, and in loving memory, that is how I wish to dedicate what follows.

Contents

List of Illustrations

Except where indicated, quotations from Shakespeare are taken from *The Complete Works*, ed. by John Jowett, William Montgomery, Gary Taylor, and Stanley Wells, 2nd edn (Oxford: Oxford University Press, 2005). In quotations from early modern editions, spelling has been modernized, but changes to punctuation have been kept to a minimum (i.e., where they are an aid to clarity). In endnotes I have retained original spelling of book titles to aid searching in online databases.

A book on Shakespeare and science has some explaining to do. For one thing, science is probably not a topic that comes as readily to mind in relation to Shakespeare's works as it does with, say, those of Mary Shelley or H. G. Wells. Even among early modern writers, Shakespeare is not so obviously preoccupied with science as some of his fellows: take Christopher Marlowe, for instance, the protagonist of whose *Doctor Faustus* can be regarded as a prototypical 'mad scientist' committing a terrible crime in search of secret knowledge; or John Donne, whose poetry explicitly registers the challenge posed by 'new philosophy' to traditional ways of viewing the universe; or, later, Margaret Cavendish, whose *Blazing World* of 1666 is prime candidate for the first English science-fiction novel.[1] The worlds of Shakespeare's plays, by contrast, are noticeably hospitable to the numinous or supernatural, whether in the form of prophecies like those uttered in *Richard III* or *The Winter's Tale*, fairies and spirits as in *A Midsummer Night's Dream* or *The Tempest*, or ghosts like those of Julius Caesar or Banquo. Whereas for Donne the 'new philosophy calls all in doubt', Hamlet's encounter with his father's ghost instead prompts scepticism about the possibility of explaining the universe through rational enquiry: 'There are more things in heaven and earth, Horatio, | Than are dreamt of in our philosophy' (1.5.169–170).[2]

Nor do the positions Shakespeare tends to occupy in the institutional and cultural structures of the modern world encourage a linkage of him with the scientific. In his Rede Lecture of 1959, which famously identified the persistence in British society of 'two cultures', 'Literary intellectuals at one pole—at the other scientists', C. P. Snow used him as the base line of familiarity with literary culture:

A good many times I have been present at gatherings of people who, by the standards of the traditional culture, are thought highly educated and who have with considerable gusto been expressing their incredulity at the illiteracy of scientists. Once or twice I have been provoked and have asked the company how many of them could describe the Second Law of Thermodynamics. The response was cold: it was also negative. Yet I was asking something which is about the scientific equivalent of: *Have you read a work of Shakespeare's?*[3]

From the perspective of the 2020s, when the UK government is reducing funding for arts and creative subjects in English universities in favour of 'high-cost subjects, including science, technology, engineering and mathematics (Stem), medicine and healthcare', that it regards as 'particularly important to the country and economy', Snow's 'two cultures' analysis still seems pertinent, albeit now in a context where it can seem that it is literary culture that has to defend its relevance.[4] This makes Shakespeare's paradigmatic status all the more stark: as England's national poet, the only writer stipulated as compulsory in the English National Curriculum, and the presiding genius of institutions like the Royal Shakespeare Company and the Globe, Shakespeare enjoys a degree of political and cultural endorsement denied other figures in an embattled arts and humanities sector, perhaps never more obviously than during the succession of stagings, adaptations, performances of new work, exhibitions, concerts, and walks occasioned in 2016 by the four-hundredth anniversary of his death.[5] If any one figure can be regarded as embodying the literary humanities in an English context (and, to lesser extents, in Britain and the English-speaking world), that figure is Shakespeare.

There are some obvious responses to these hypothetical caveats. Clearly, neither Shakespeare's status as figurehead for the humanities nor the depiction of ghosts, fairies, and the like within his work precludes his engaging with scientific themes. The conceptual distinctions between science, magic, and religion that are familiar and obvious-seeming in the modern world were much less so in Shakespeare's time, as is indicated by the fact that when Marlowe's Faustus engages in detailed discussion of the movement of heavenly bodies it is with a devil to whom he has sold his soul in return for magical powers. However, that very conceptual indistinctness suggests a third objection that is somewhat more substantial, and that relates to the historical context within which Shakespeare was writing. Spanning the period from the early 1590s to the early 1610s, his literary career predates some of the events and discoveries that are foundational to modern science, including William Harvey's publication of his theory regarding circulation of the blood in 1628, Galileo Galilei's trial by the Inquisition following the publication of his *Dialogue Concerning the Two Chief World Systems* in 1632, the institutionalization of experimental scientific method in England with the foundation of the Royal Society in 1660, and the publication of Isaac Newton's *Principia Mathematica* in 1687. His works reflect assumptions about the nature of the universe and about human physiology that the so-called 'scientific revolution' would discard; indeed, they date from a period when the word 'science' itself had not acquired its modern meaning, and when the use of other terms to describe the study of nature reflected different assumptions about the aims and scope of that enterprise. It is therefore appropriate to ask whether a study of Shakespeare and science is fundamentally anachronistic in its terminology and outlook.

The remainder of this introduction will explore that objection more fully. In doing so it seeks to achieve a more nuanced contextualization of Shakespeare in relation to the cultures of scientific enquiry prevalent in his day, and to set out some of the ways in which this book will suggest his work reflects, responds to, and participates in those cultures. It will summarize some of the recent lines of critical enquiry into Shakespeare and science upon which it draws. And it will briefly set out the content of the chapters that follow.

Science and Natural Philosophy

On the few occasions when Shakespeare uses the word 'science', the apparent meaning is close enough to the modern sense to be recognizable, but far enough from it to be mildly estranging. In the final scene of *All's Well That Ends Well* the King of France, demanding to know how Bertram obtained the ring that he himself gave to Helen, says that even Plutus, the classical god of riches, 'Hath not in nature's mystery more science | Than I have in this ring' (5.3.104–105); here 'science' seems to mean the kind of knowledge that comes from familiarity, as it does at the start of *Measure for Measure* when the Duke tells Escalus that he declines 'Of government the properties to unfold' since Escalus's 'own science | Exceeds in that the lists of all advice | My strength can give you' (1.1.1, 3–5). When the word is used in its plural form its meaning gets a bit more specific, as when in *The Taming of the Shrew* Petruccio presents the disguised Hortensio to

Katherina's father Baptista claiming that, being 'Cunning in music and the mathematics', he can 'instruct her fully in those sciences' (2.1.56–57). Similarly, at the end of *Henry V* the Duke of Burgundy laments that following the wars between France and England 'our houses and ourselves and children | Have lost, or do not learn for want of time, | The sciences that should become our country, | But grow like savages' (5.2.56–59). In both cases the word 'sciences' indicates specific branches of learning or skill, albeit more broadly understood than in a modern context; indeed, given that a few lines earlier Burgundy has described peace as the 'Dear nurse of arts' (35), there seems to have been a good deal of overlap between what 'arts' and 'sciences' could be understood to refer to, in contrast to present-day practice. Strictly speaking, though, medieval learning had differentiated between *scientia*—knowledge of things that are unchanging, necessary, and eternal, according to the definition provided by the ancient Greek philosopher Aristotle in the sixth book of his *Ethics*—and *ars*, or 'a rational quality, concerned with making'.[6] This kind of distinction, between theoretical sciences and practical arts, persisted into Shakespeare's own day, informing the polymath John Dee's explanation that from the 'Science of *Geometry*' are derived '*Methodical Arts*' such as perspective, astronomy, and music (although confusingly, both astronomy and music were frequently described as sciences in other contexts).[7] In the Middle Ages, *scientia* could comprise branches of knowledge that we might not now think of as scientific: most strikingly, theology, which Thomas Aquinas argued 'from every standpoint ... excels all other sciences' on the grounds of its subject matter, its dependence on God's knowledge rather than fallible human reason, and its end goal of salvation.[8]

One term still used in Shakespeare's time that better corresponds to 'science' in its modern sense is 'natural philosophy', which 'studied the material world as it was visible to the senses', seeking 'the universal causes of phenomena' (in contrast to natural history, which concerned itself with particulars).[9] In the context of medieval universities this generally entailed the study of relevant works by Aristotle such as the *Physics*, *On the Heavens*, and *Meteorology*.[10] Although the term would later be used by figures now seen as foundational for modern science (for instance, by Robert Boyle in the title of his *Considerations Touching the Usefulness of Experimental Natural Philosophy*, 1663), it is still not quite synonymous with 'science', not least because of its emphasis on providing 'causes for change in nature' relating to topics such as 'the elements and transformations of them, the growth, decay, and classification of living things, motion, the heavens, and so forth'.[11] In the context of medieval Christianity the justification for this enquiry into causes was expressed in religious terms: the study of nature was undertaken with the aim of understanding the mind and intention of God who created it. This did not preclude either the minute observation of natural phenomena or the production of sophisticated technology, but it did provide natural philosophy with a different conceptual underpinning from that of modern science.[12] Shakespeare demonstrates familiarity with 'natural philosophy' as a term when he sets up a dialogue between the sardonic Touchstone and the idealized rustic Corin in *As You Like It*:

> TOUCHSTONE   Hast any philosophy in thee, shepherd?
> CORIN   No more but that I know the more one
>   sickens, the worse at ease he is, and that he that
>   wants money, means, and content is without three
>   good friends; that the property of rain is to wet, and
>   fire to burn; that good pasture makes fat sheep; and
>   that a great cause of the night is lack of the sun; that
>   he that hath learned no wit by nature nor art may
>   complain of good breeding, or comes of a very dull
>   kindred.
> TOUCHSTONE   Such a one is a natural philosopher.
>   (3.2.21–31)

The tone here is hard to gauge: Touchstone punningly insinuates that Corin is a 'natural' or fool, but Corin's responses are entirely in keeping with his pastoral simplicity, making Touchstone's condescension seem rather misplaced. It is also difficult to be sure how much of the satirical intent is directed against natural philosophy itself, given the way in which Corin's seemingly circular responses recall some of Aristotle's primary statements about what the universe is made of: 'Fire is hot and dry, Air is hot and moist …, Water is cold and moist, and Earth is cold and dry.'[13]

A more explicit scepticism about established ways of studying nature is evident in the writings of Francis Bacon, whose views on this topic would prove highly influential for the development of scientific thinking in the seventeenth century. Bacon writes in the fifteenth of his 'aphorisms of the interpretation of nature and kingdom of man' in the *Novum Organum* (1620; translated 1676) that concepts like 'heavy, light, thick, thin, moist, dry, generation, corruption, to attract, to expel, element, matter, form, etc'. are not 'proper Notions' but 'phantastical and ill designed'; the basic ideas about nature expressed in terms like these have no basis in reality but are mere 'Idols or false Images, which besiege Men's minds'.[14] Bacon's comment exemplifies what some historians have seen as a decisive shift in early modern scientific thinking away from the presumption that nature was legible in human terms and towards the belief that the way the universe appeared to the senses might be misleading. In her study of the impact of 'new science' on the literature of the period, Mary Thomas Crane singles out 'Copernican astronomy, advances in mathematics and optics, particulate theories of matter, and germ theories of disease' as emergent models of how nature worked that offered the troubling implication that 'lived experience of the world did not reliably provide access to truth about it'. She sees Shakespeare and his contemporaries as living in a period of uneasy transition, when 'settled … accounts of how the universe worked began to fall apart and the new ideas that would replace them were still inchoate and in flux'.[15]

A 'Scientific Revolution'?

While Crane's account focuses specifically on a group of discrete scientific theories and their discernible impact on early modern English literature, some historians of science have argued that new ways of conceptualizing, studying, and working with nature during this period were more far-reaching in their effects and implications. An influential example is Herbert Butterfield in his 1948 lectures for the History of Science Committee in Cambridge, later printed as *The Origins of Modern Science*:

the supremely important field for the ordinary purposes of education is … the so-called 'scientific revolution,' popularly associated with the sixteenth and seventeenth centuries, but reaching back in an unmistakably continuous line to a period much earlier still. Since that revolution overturned the authority in science not only of the middle ages but of the ancient world—since it ended not only in the eclipse of scholastic philosophy but in the destruction of Aristotelian physics—it outshines everything since the rise of Christianity and reduces the Renaissance and Reformation to the rank of mere episodes, mere internal displacements, within the system of medieval Christendom.

The topics surveyed in Butterfield's account include ones already referred to in this introduction, such as the model of the universe proposed by Copernicus whereby the Earth orbited the Sun rather than vice versa, Harvey's theory of circulation, and the rise of experimental method. For him, rather than being a collection of unrelated developments they amount to something even more substantial: the wholesale rejection of an existing orthodoxy that stretched back to the ancient world and was epitomized in the figure of Aristotle. This revolution, Butterfield argues, transformed 'the whole diagram of the physical universe and the very texture of human life itself', and constitutes 'the real origin both of the modern world and of the modern mentality'.[16] Butterfield was not alone in using the term 'scientific revolution' to denote this alleged phenomenon: Alexandre Koyré's article on 'Galileo and the Scientific Revolution of the Seventeenth Century' had appeared in 1943, and in the postwar decades the term became increasingly prevalent, as in titles like A. Rupert Hall's *The Scientific Revolution, 1500–1800* and J. D. Bernal's volume on *The Scientific and Industrial Revolutions* (both 1954).[17]

While Butterfield does treat the sixteenth century as his starting point, as well as gesturing back to earlier periods, the way he defines the scientific revolution raises the question posed earlier in this introduction of how one relates Shakespeare to the advances he mentions. Shakespeare lived too early to be affected by the discoveries of Galileo or Harvey, while even sixteenth-century developments such as Copernicus's theory of heliocentrism are at best ambiguously present in his works. Indeed, his writings explicitly rely on models that science would come to discard, such as when Menenius's fable of the belly in *Coriolanus* imagines the human bloodstream not as a single circulatory system driven by the heart, but as something more akin to a network of channels via which the various organs actively propel organic materials to one another. The belly explains its role in the distribution of digested food as follows:

> I send it through the rivers of your blood
> Even to the court, the heart, to th' seat o'th'
> brain;
> And through the cranks and offices of man
> The strongest nerves and small inferior veins
> From me receive that natural competency
> Whereby they live. (1.1.133–138)

Menenius's description serves reasonably well as an illustration of the unequal model of wealth distribution that he is trying to justify. After Harvey, though, it loses validity as an account of human biology, inviting the question: can literature based on incorrect assumptions about nature be called scientific?

A pragmatic answer might be that while Shakespeare's writing predates a number of the changes in understanding that are central to the scientific revolution as Butterfield (for example) conceives it, he would nevertheless have been aware of multiple challenges to medieval and ancient orthodoxy. These included discoveries in the field of anatomy by Andreas Vesalius and others, which disproved long-held assumptions about human biology (see Chapter 3), as well as voyages of discovery and conquest that encountered continents of which the ancients had been unaware and which showed the 'torrid zone' around the equator to be habitable even though, as a speaker in Robert Recorde's *Castle of Knowledge* puts it, 'all the old Cosmographers did think that that country might not be inhabited for heat'.[18] As Adam Max Cohen has emphasized, the London in which Shakespeare had his career was undergoing 'a multifaceted technology boom', while Deborah Harkness has vividly set out the networks of naturalists, physicians, instrument-makers, and chemists at work in it.[19] In addition, since well over a century before his birth ancient texts that had largely been unavailable to medieval Europe, from the mystical *Corpus Hermeticum* to Lucretius's poem *De rerum naturae* (which describes the atomist philosophy of Epicurus), had been rediscovered and given renewed circulation, assisted by the new invention of printing. These offered precedents for contemporary models of nature that broke with accepted beliefs, while the products and cultures of early modern London—as of other major cities—helped facilitate the processes of observation and enquiry that would sustain scientific investigation over future decades. Shakespeare's age can be seen as the product of a period of profound change in how humans were able to view and act on nature, as well as providing a foundation for more far-reaching transformation in the century that followed. His inevitable unawareness of future discoveries, and his participation in the errors of his own time, do not negate the ways in which new challenges to old orthodoxies were able to inform his work.

There is, however, a different way of responding to the objection that the time when Shakespeare wrote and the factual errors he makes render his work inherently unscientific. That is to question—as historians of science increasingly do—the narrative of an early modern 'scientific revolution' that vanquished medieval misconceptions, replaced them with more accurate models of the structure and processes of the universe, and established a firm methodological basis for future scientific investigation. Steven Shapin, for example, expresses scepticism as to whether 'there was any single coherent cultural entity called "science" in the seventeenth century to undergo revolutionary change', instead positing 'a diverse array of cultural practices aimed at understanding, explaining, and controlling the natural world, each with different characteristics and each experiencing different modes of change'. He is similarly dubious as to whether any such change can be dated straightforwardly to the seventeenth century, and about the extent to which it might be deemed revolutionary.[20] Some practitioners of and writers on science in the period certainly did present the sixteenth and seventeenth centuries as having seen a significant advancement upon former times: witness Thomas Sprat in his 1667 history of the Royal Society, which contrasts the 'experimental philosophy' he advocates with the learning of medieval scholastics who 'were tied by their Cloistral life, to such a strictness of hours, and had seldom any larger prospects of *Nature*, than the Gardens of their *Monasteries*'. Sprat, an Anglican Bishop, presents his own time as the '*third great Age* of the *flourishing* of *Learning*' after those of the Greeks and Romans, insisting that 'If we compare this *Age* of *Learning*, with the *two former*; we shall find, that this does far exceed both the other in its extent: there being a much larger plot of ground, sown with Arts, and civility at this time'. However, his remarks need to be read in relation to the text in which they appear: the history of an organization for the 'Advancement of Experimental Philosophy', only recently founded, whose practices Sprat evidently felt required some justification.[21] His strategy here contrasts strikingly with those of earlier figures such as Copernicus (who pointed to the Greek astronomer Aristarchus as a precedent for his heliocentric theory) and Vesalius (who insisted on his veneration for the Greek anatomist Galen, despite the multiple respects in which his findings proved Galen wrong).[22] Evidently, an emphasis on novelty or an emphasis on tradition could be rhetorically useful for early modern scientists, depending on the context.

One factor that should complicate a simplistic view of medieval ignorance and superstition giving way to evidence-based science is the fact that, as revisionist historians often note, several major figures held beliefs that are no longer considered scientific today. Obvious examples are Isaac Newton's belief in alchemy and his attempts to decipher prophetic messages in the Book of Daniel, and the astronomer Johannes Kepler's belief that the orbits of planets around the sun corresponded to the relative dimensions of the five Platonic solids, reflecting the divinely planned harmony of the universe.[23] Conversely, precursors for major epistemological shifts alleged to characterize the scientific revolution have been found in the work of medieval scholars: Edward Grant, for example, has argued that the speculations of medieval theologians about extracosmic void space 'helped shape … spatial interpretations' in seventeenth-century physics.[24] A recent assertion of the importance of the Middle Ages in the history of science, Seb Falk's *The Light Ages*, offers a powerful argument for the sophistication of medieval scientific practice, especially as exemplified in the observation and prediction of celestial movements and in the construction of devices that applied these observations to the accurate recording of time such as the astronomical clock invented by Richard of Wallingford and housed at St Albans abbey.[25] The role of Islamic cultures in developing techniques and technologies of astronomical measurement (not to mention other branches of learning) that would be taken up in Christian Europe also needs to be emphasized, given the tendency of some European historians to diminish it: Sprat represents an early example of this tendency, admiring the '*Arabians*' as 'men of a deep, and subtle Wit' but asserting that their achievement 'mainly consisted, in understanding the Ancients' since their work 'was not fully over, before they were darkened by that, which made even *Greece* itself Barbarous, the *Turkish Monarchy*'.[26]

Other objections to or qualifications of the notion of a scientific revolution can be offered, not least that it overstates the reach and significance of work done by a minority of educated males in a small part of the world.[27] However, as far as the reciprocal relevance of Shakespeare and science is concerned, the main point is this: once the conceptual validity of the scientific revolution is called into question, whether on the grounds of over-simplification, the self-interestedness of its proponents, its marginalization of medieval thought, its Euro-centrism, or for other reasons, its legitimacy as a basis for marking out what does and what does not get to be considered scientific is eroded. What Mary Lindemann characterizes as 'a version of history that postulates a single and relatively straightforward passage into the modern world leading from the dark ages of ignorance, superstition, and suffering into a brighter world of knowledge, science, and abundance' can be rejected in favour of a mode of enquiry that is more ready to see scientific knowledge, whether in the past or the present, as the product of complex, contradictory, and socially based structures, not the triumphant unveiling of pre-existing truths.[28] This focus on the contexts in which scientific knowledge is made is a central feature of science studies, a field of enquiry that in David J. Hess's words 'tracks the history of disciplines, the dynamics of science as a social institution, and the philosophical basis for scientific knowledge'.[29] It also informs the way the goal of the history of science is articulated by Peter Dear: 'to understand why particular people in the past believed the things they did about the world and pursued enquiries in the ways they did. The historian has no stake in adjudicating the truth of past convictions'.[30] For the purposes of literary enquiry, this means that a study of 'Shakespeare and science' can include both a discussion of the ways in which Shakespeare registered developments in the study of nature that future ages would assimilate to their model of scientific progress, and an account of how his writing was informed by venerable (and, to put it bluntly, inaccurate) assumptions about the universe, matter, and the body. As attempts to provide rational explanations for natural phenomena, these can legitimately be described as science rather than as false starts or mistakes that 'real' science would come to reject or supersede—as if those two categories could be distinguished so straightforwardly. Models accepted as scientifically valid in one time often bear the imprint of older systems whose premises they have discarded, as Ofer Gal argues using the example of Freud's notion of hysteria:

We have the 'unconscious' because Sigmund Freud was intrigued and fascinated by women who were blind despite healthy eyes or paralyzed despite healthy limbs. But we no longer have hysteria, the malady that Freud explained psychologically…. Yet 'unconscious' is still as integral a part of psychology as it is of general culture.[31]

Another example more relevant to Shakespeare's age might be the persistence in modern science of terms like 'atom' (see Chapter 2) and 'temperature' (see Chapter 3), both derived from ancient theories within which they had somewhat different meanings from their modern ones: their continued use reflects the ancestry of new models in the old ones they ostensibly supersede.

An awareness of this interconnectedness of modern and early modern (and earlier) ways of understanding the world, accompanied by an insistence on the validity of the latter on their own terms, informs my practice throughout this book of using the word 'science' to denote knowledge systems that would not have been given that designation in Shakespeare's time and that science has since discarded. Anachronistic though this may be, it offers a way of engaging with them without condescension, and of asserting their potential relevance to our own attempts to make sense of and act upon the natural world, as well as having the more pragmatic advantage of brevity.

Shakespeare and Science

The paragraphs above have incidentally suggested some reasons for studying Shakespeare in relation to science: the way his work troubles modern binaries of the scientific versus the supernatural, and of literature versus science; the fact that he wrote during a period when dominant ideas about the universe, nature, and how to investigate them were being called into question, and on the eve of what would come to be seen as a period of major transformation in this respect (however problematically and debatably); the insights his work can provide both into old worldviews and into how new ones were received. Another, more practical reason is the need to have some familiarity with topics like Galenic medicine and Ptolemaic cosmology when trying to understand Shakespearean imagery and metaphors, as well as the conceptual underpinning of ideas such as racial and gender difference as they are treated in his works. However, these justifications are largely contextual, positioning early modern science as part of the background to Shakespeare's writing. They do not tackle the more complex question of whether Shakespeare addresses scientific ideas and themes in any more active way: whether his work goes beyond merely reflecting ideas about how the universe worked that he encountered in his reading and in the wider culture, instead subjecting them to scrutiny, creative reshaping, and other forms of artistic and intellectual engagement.

That shift of focus from reflection to engagement has been a recurrent feature of critical enquiry into early modern literature and science over recent decades. In her introduction to a special issue of the *South Central Review* devoted to 'Shakespeare and Science', Carla Mazzio notes that its constituent essays are characterized by 'a prevailing interest in moving beyond forms of analysis focused largely on thematic traces of, or indeed linguistic reflections of, historically specific arenas of scientific practice' in Shakespeare's work, attempting instead 'a more nuanced understanding of historical relationships between dramatic and scientific practices, epistemologies, and mentalities'.[32] A similar point is made by Howard Marchitello and Evelyn Tribble, who in their introduction to *The Palgrave Handbook of Early Modern Literature and Science* cite works by Charles M. Coffin, William Empson, and Marjorie Hope Nicolson as examples of an older approach:

Within the ascendency of New Criticism, these were scholars who were deeply invested in understanding early modern science as they found it reflected in the literature of the period. Their efforts were largely dedicated to reading literary writers (especially the poets, with John Donne as perhaps their most significant exemplar) against the backdrop of the new science. This work—as innovative and important and compelling as it was—was nevertheless confined to what can be considered influence study: how did English poets respond to the frequently radical and potentially destabilizing effects of scientific writing?

By contrast, as Marchitello and Tribble explain, more recent critics are increasingly preoccupied with 'the ways in which both science and literature are mutually informing and mutually sustaining. The presiding question is no longer how do literary writers react to scientific writers, but rather how are the literary and scientific practices imbricated?'[33] This approach has been informed by a growing theoretical self-consciousness, in the wake of New Historicism, about the difficulty of separating texts from contexts, and an accompanying reluctance to position early modern history and culture simply as backgrounds to the privileged art objects in the foreground. It involves treating literary texts not as passive reflectors of scientific discourses, but as both permeated by them and responding to them in more complex ways; and, at the same time, recognizing the extent to which 'the impact of natural philosophy on the early modern intellectual landscape depended on the effective use of rhetorical and even dramatic techniques to communicate and develop new ideas', as Juliet Cummins and David Burchell put it.[34]

The presence of literary qualities in early modern scientific texts is testimony to the fact that literature and science were not regarded as separate fields during that period to the extent that they are today. Defending poetry against its detractors as 'the first light-giver to ignorance', Philip Sidney noted that 'the philosophers of Greece durst not a long time appear to the world but under the masks of poets. So Thales, Empedocles, and Parmenides sang their natural philosophy in verses.'[35] The ancient assumption that poetry was an appropriate medium for the exposition of scientific material persisted into Shakespeare's time and beyond: the physician Girolamo Fracastoro set out his theories of disease in the poem *Syphilis*, while the influential *Zodiacus Vitae* of Marcellus Palingenius Stellatus includes a detailed account of the nature of the cosmos. Scientists such as Pierre Gassendi and Isaac Newton who were influenced by the atomist philosophy of Epicurus found it most fully developed in the poetry of Lucretius, as will be discussed in Chapter 2, and they had no qualms about engaging with Lucretius directly. The idea that scientific writing needed to distance itself from poetry only became widely prevalent in the seventeenth century, when Sprat's *History* would praise the founders of the Royal Society for having 'endeavoured, to separate the knowledge of *Nature*, from the colours of *Rhetorick*, the devices of *Fancy*, or the delightful deceit of *Fables*'.[36]

A recurrent theme of this book will be the extent to which Shakespeare's thinking about science is informed by texts and discourses that one might not immediately think of as scientific, including the *Metamorphoses* of Ovid, the essays of Montaigne, and religious controversy over the Eucharist. At the same time, Shakespeare was demonstrably aware of Aristotelian, medieval, and contemporary ideas about nature and the universe to an extent that indicates wide reading in more specialist literature. His lack of a university education was no barrier in this respect: as will become apparent over the chapters that follow, an extraordinary amount of material was available in print, and in English, that brought scientific information about the cosmos, the human body, and other topics to a general readership. (Indeed, this limitation may have been a positive advantage: as Francis R. Johnson notes with respect to astronomy, 'the better textbooks in the vernacular were more up-to-date, and superior in scope and quality to those commonly used in the Schools'.)[37] The uses to which Shakespeare puts these materials are diverse. They range from the deployment in his history plays of current theories about how lightning, shooting stars, and earthquakes were produced (see Chapter 1), through a detailed consideration of the physiology of emotional response in *Lucrece* (Chapter 3), to the reapplication in *Henry V* of advice about how to do calculations with Hindu-Arabic numerals (Chapter 4). However, as well as highlighting such instances, this book will argue for a more critical engagement with scientific themes on Shakespeare's part. In Chapter 1, for instance, I show how the genre of the history play, both for Shakespeare and for other dramatists, was informed at a deep level by a tension between religious and natural explanations both for meteorological phenomena and for political outcomes; as well as embodying this tension, Shakespeare's plays critique the way humans understand and exploit it. In Chapter 2, I argue that through the figure of Prospero in *The Tempest* Shakespeare scrutinizes the claims of early modern science to be able to achieve mastery over nature. And my discussion of humoral theory in Chapter 3 shows how Shakespeare both draws on it as a means of exploring changing emotional states and depicts characters who use it as a means of rationalizing their own and others' behaviour. His references to the culture of anatomical demonstration that had already posed challenges to Galenic orthodoxy register the fascination that the prospect of opening the body's interior to view exerted over the early modern imagination, but also scepticism over the capacity even of such intrusive methods of investigation to serve as a gateway to truth.

The way my analysis has been split across the four chapters is not intended to reflect any equivalent systematization of scientific enquiry in the early modern period. As will become apparent, fields that modern science might tend to separate were differently understood in Shakespeare's day, when, for example, the supposed influence of heavenly bodies on the progress of disease meant that a physician was expected to have a good understanding of astronomy.[38] To give another example, the atomist theories advanced by Lucretius were underpinned by a belief in the universe as infinite that contradicted Aristotelian, and medieval Christian, orthodoxy: thus, different ways of thinking about matter implied different ways of thinking about the cosmos. The pragmatic need to find a way of dealing with these interrelated materials (however unsatisfactory) leads me to divide the book into chapters that reflect four different areas of scientific interest or practice. I begin in Chapter 1 by considering the broadest possible context for the subjects addressed in later chapter, namely the cosmos, the movements of heavenly bodies, and the other more local meteorological events that these bodies were understood to influence. In the early modern period, as at other moments in history, such events were variously interpreted as divine warnings or punishments and as mere natural phenomena (albeit within a divinely created universe); accordingly, this chapter deals with themes of causality, of epistemology, and of the proper scope of scientific investigation itself that are foundational for the rest of the book. The chapter that follows comes back down to earth: specifically, to the materials from which terrestrial substances were made. It argues that even within individual texts, multiple ways of conceptualizing the nature and behaviour of matter can be identified: in the Sonnets, for example, the familiar model of the four elements coincides with ideas drawn from alchemy and with terms elsewhere used to describe what happens to bread and wine in the Eucharist. Much of the analysis in this chapter focuses on tragedy, as a dramatic genre inherently concerned with mortality and the reduction of humans to inert matter; however, the chapter concludes with *The Tempest*, a play that seems to depict both the transformation of matter and its rarefaction into 'thin air'. Chapter 3 considers the human body not as mere matter but as biological system, juxtaposing the assumptions about it that Shakespeare derived from centuries-old models with the findings of early modern anatomy—a form of scientific enquiry that fascinated Shakespeare and his contemporaries. I begin with an examination of *Lucrece*, a poem that reflects early modern assumptions about the relationship between mind and body in its emphasis on the embodied nature of emotional response. I then discuss Shakespeare's treatment of humoral physiology in *Henry IV* and *Othello*, before analysing the ways in which both *Othello* and *Hamlet* interrogate anatomy's claim to be able to generate bodily knowledge. Finally, in Chapter 4 I consider mathematics, a way of representing and working with nature that enabled advances in all of the fields discussed in the earlier chapters. As a property owner and shareholder, Shakespeare was evidently familiar with the use of mathematics in accounting calculations. However, a particular stimulus to his creative imagination appears to have come from Hindu-Arabic numerals themselves, with their shifting signification and their susceptibility to the power of the zero to multiply or to negate—attributes that Shakespeare explores in *Henry V* and *King Lear*.

My discussion of Shakespeare and science is deeply indebted to the many critics and historians who have broken new ground in this area over recent decades. Throughout the chapters that follow, I have tried to do justice to their work, both in the interests of acknowledging my debts to it and in order to provide others researching the topic with signposts towards some key texts. Thus, in Chapter 1 I locate my discussion of astronomy and meteorology in the history plays and *Hamlet* in relation to recent scholarship on the epistemological anxieties generated by new ideas about the place of the Earth in the cosmos, as well as work done by Gwilym Jones and Sophie Chiari on the meanings attached to weather in the early modern period. In Chapter 2 I address the question of how matter and materiality are conceptualized in Shakespeare's tragedies in light of current debates over the extent of his familiarity with atomist concepts; my subsequent discussion of *The Tempest*, a play that depicts in Prospero a character who has gained control over the elements, draws both on work that troubles the distinction between Renaissance magic and science, and on

Denise Albanese's research into the discursive links between scientific enquiry and colonization. My discussion in Chapter 3 of how the body is imagined in Shakespeare is informed by the extensive analysis of Gail Kern Paster, Michael Schoenfeldt, Erin Sullivan, and others of the influence of humoralism on his writing, and by the insights of scholars such as Kim F. Hall into how humoral theory was used to articulate ideas of racial difference. Finally, the topic with which I engage in Chapter 4 is one that has garnered increasing interest over recent years (as from Paula Blank, Joseph Jarrett, Shankar Raman, and Henry S. Turner), namely the intersection of literary and mathematical discourses—a theme I take up with particular focus on *Henry V* and *King Lear*. I am acutely aware that the brevity of this book, the vastness of the subject, and my own scholarly limitations mean that my critical survey can only be a partial one, and I hope that both readers and authors will forgive me the many omissions.

In addition to the more localized arguments about specific texts and topics that I put forward in the individual chapters, I aim to offer several broader contributions to the subject. First, I have tried under the heading of 'Shakespeare and Science' to bring together a multiplicity of fields in which invaluable work has been and continues to be done, including not only the areas of science I focus on (astronomy and meteorology; theories of matter; the body; mathematics) but also ecocriticism, religion, the reception of classical literature, feminist criticism, history of the emotions, and race and racialization, to give just a few examples. Although my discussion of all of these topics is less in-depth than would be possible in a more narrowly focused book, I hope to have provided a sense of the great diversity of ways in which Shakespeare uses and critiques the knowledge about the natural world that was available to him.

This leads on to my second contribution, which is the argument that—notwithstanding the caveats adumbrated at this start of this chapter—Shakespeare as a writer had an omnivorous interest in science, from the workings of the cosmos to those of the human body, right down to the 'many a thousand grains' (*Measure for Measure*, 3.1.20) from which it was constituted. This may have been the result of personal disposition, but it can also be attributed to the need to furnish theatregoers with striking images, bold rhetoric, and lively characterization, a purpose to which his use of scientific images and ideas is frequently applied. One other, broader factor may have been poetic ambition. Classical poets like Lucretius, Virgil, and Ovid had all included in their great works a good deal of the scientific knowledge of their day, as with the discussion of astronomy and meteorology in the first book of Virgil's *Georgics* or the vision of the constant processes of change at work in nature in the final book of Ovid's *Metamorphoses*. Shakespeare's discussion of similar topics can be read as an attempt to claim a comparable stature for himself as a poet, in line with the view of him as self-consciously maintaining 'a dialogue with such poets from antiquity as Virgil and Ovid' that has been advanced by Patrick Cheney.[39]

At the same time, though, the fact that so much of Shakespeare's work takes the form of drama gives its engagement with scientific materials a fundamentally different character from that in the works cited above. Drama is an inherently dialogic form, one that relies on multiple speakers bringing their diverse understandings of the world into contact and, potentially, conflict. This allows Shakespeare's plays to be open to different, even contradictory, models of nature. In contrast with the worked-out cosmologies of Dante's *Divine Comedy* or Milton's *Paradise Lost*, Shakespeare's plays do not offer a consistent and coherent account of the universe and the things in it: as Katherine Walker has recently noted, 'There is not a unified paradigm of science, nor of nature ... in all of Shakespeare's works'.[40] From one perspective, this could be seen as making them (again) fundamentally unscientific; from another, it leaves them generatively open and provisional, allowing for the juxtaposition of opinions such as those of the Aristotelian Hotspur and the 'great magician, damned Glyndŵr' in *1 Henry IV* (1.3.82). They can comprehend Hamlet's stress on the obdurate persistence of matter, which generates the notion that the dust of Alexander the Great might be stopping a beer-barrel somewhere, and Prospero's sense of the elements as both manipulable and transient, things that can 'dissolve; | And ... Leave not a rack behind' (*The Tempest*, 4.1.153–156). In their treatment of scientific themes Shakespeare's texts are places in which to think through possibilities, to work out hypotheses, rather than expressions of a settled opinion.

Finally, where possible I have tried in this book to be alert not just to the dialogic nature of Shakespeare's dramatic writing, but to its theatricality. Recent work on the history of science has asserted the importance, not just of practitioners and their networks, but of the spaces in which scientific work was carried out. The *Cambridge History of Science* volume on early modern science edited by Katharine Park and Lorraine Daston includes a number of chapters devoted to 'sites of natural knowledge', which extend from obvious examples such as anatomy theatres and laboratories to 'Markets, Piazzas and Villages', 'Homes and Households', and 'Coffeehouses and Print Shops'. It is legitimate to consider whether the early modern playhouse represented another such site, offering a visual and spatial representation of natural phenomena and, as Henry S. Turner puts it, a place 'for modelling large-scale social processes'.[41] The chapters that follow draw on the work of scholars such as Kristen Poole, who likens the Globe playhouse to an observatory; William West, who links the performance practices of early modern actors to theories of matter; and Jonathan Sawday, who discusses the affinities between playhouses and anatomy theatres. Insights such as these underline the extent to which theatres themselves were environments where an understanding of science was generated through performance, illusion, and the creative use of space. As Jean Feerick points out, the name of Shakespeare's Globe identified the playhouse as a 'mini earth', proclaiming it an appropriate setting in which to stage 'debates about matter, nature, and the cosmos'.[42] However, the Globe also shared its name with a scientific instrument whose manufacture, as Cohen shows, was becoming increasingly advanced in early modern London, and which Recorde described as 'the ground and beginner of all other instruments'.[43] Just as the title of Shakespeare's earlier venue the Theatre had aligned it with the dramatic culture of the ancient world (probably more so than with the 'anatomy theatre', a term just coming into use in the period), this choice of nomenclature made a claim for the playhouse's status and significance: in this case, about its capacity to represent both the wider world and the (supposedly) spherical heavens beyond.[44] Far from being structurally opposed to or incompatible with scientific enquiry as a 'two cultures' model of science and the arts might suggest, Shakespeare's theatres were well positioned not only to reflect but to reuse, to remodel, and to critique the scientific ideas current in their time.

1

Things in Heaven and Earth

Midway through *Doctor Faustus* (*c.* 1588–92), Christopher Marlowe's play about a man who sells his soul in the hope of obtaining knowledge and power, the Chorus summarizes Faustus's actions since we last saw him:

> Learnèd Faustus,
> To know the secrets of astronomy
> Graven in the book of Jove's high firmament,
> Did mount himself to scale Olympus' top,
> Being seated in a chariot burning bright,
> Drawn by the strength of yoky dragons' necks.
> He now is gone to prove cosmography ...[1]

In order to 'know the secrets of astronomy', Faustus needs a flying machine pulled by dragons, and this brings home the extent to which, when Marlowe was writing, the astronomical knowledge that could be obtained from Earth was inherently limited and hypothetical. Nearly two decades after Marlowe's death, and in the last years of Shakespeare's life, the Italian astronomer Galileo Galilei would use a telescope—a recent invention, upon which he had improved—to observe the surface of the Moon and to identify phenomena such as the moons of Jupiter and the phases of Venus. These discoveries, which he made public in his *Sidereus nuncius* or *Starry Messenger* in 1610 and in his 1613 work on sunspots, were of epochal importance because they disproved prevailing assumptions about the nature of the universe, in particular the idea that the Earth was the centre around which the heavens revolved. A heliocentric or sun-centred model of

Shakespeare and Science. Tom Rutter, Oxford University Press. © Tom Rutter (2024).
DOI: 10.1093/oso/9780192898548.003.0002

planetary movement had been proposed decades earlier, by Nicolaus Copernicus in his *De revolutionibus* of 1543, and Copernicus himself cited the ancient astronomer Aristarchus as an earlier adherent of the same view. However, this model could be no more than a hypothesis until Galileo was able to observe the planets in unprecedented detail by means of a telescope: to 'prove' it, in the words of Marlowe's Chorus.

It is useful to bear in mind Marlowe's description of astronomical knowledge as 'secret' when considering the subject of this chapter, namely the understanding of the universe, as well as of more local meteorological phenomena, that informed Shakespeare's writing. By Shakespeare's era astronomy was a well-developed science whose practitioners were able to make accurate predictions about the movements of heavenly bodies for years into the future, while writers on the weather offered theories about the causes of clouds, rainbows, and the like that have elements in common with the views of modern meteorologists. At the same time, their arguments were based on hypotheses about the structure of the universe, the relationship between bodies below the Moon and ones beyond it, and the causes of weather events that were in large part unverifiable and have since been disproved. An uneasy awareness that traditional explanations may be unsatisfactory is evident in *Hamlet*, and this chapter will end with a discussion of that play's treatment of astronomy. However, it will also propose that Shakespeare's treatment of astronomy and meteorology is of interest not simply in terms of its awareness of recent developments, but also for its focus on the human dimension of these fields of enquiry: how people make sense of the things above them, and how they use that understanding for their own purposes. These questions are particularly relevant to the first set of texts I will discuss in detail, namely Shakespeare's history plays of the 1590s, which are driven by a tension between a view of both nature and human history as being shaped by divine will and an interest in secular explanations. As I shall argue, this tension was partly inherited from earlier history plays, which along with more obviously meteorological texts represented an important influence on Shakespeare's thinking and practice. Before that, this chapter will summarize some features of Renaissance thinking on astronomy and meteorology that are particularly relevant to Shakespeare's plays, and survey some relevant criticism.

The Earth in the Universe

As a literate Englishman in the late sixteenth century, Shakespeare was able to draw on a wide variety of sources for information about the universe and the Earth's place within it. One example is the highly popular *Calendar of Shepherds*, a regularly reprinted compendium of useful information on diverse topics including the dates of religious festivals, the nature of virtues and vices, human anatomy and health, and astrological prediction. It includes detailed descriptions of the structure of the heavens, the divisions of the night sky, and the constellations, geared towards practical considerations such as telling the time, understanding the effects of the stars on human fortunes, and knowing when to undertake or avoid particular activities. The sixteenth century also saw an influx of English translations and adaptations of more high-status texts such as Stephen Batman's revised version (1582) of the encyclopaedic *De proprietatibus rerum* (*c.* 1240), by Bartolomaeus Anglicus, and Barnabe Googe's 1560s translation of the *Zodiacus vitae* (1536), by Marcellus Palingenius. The latter of these, which renders the hexameters of Palingenius into English fourteeners, is a good example of the extent to which poetry was considered a suitable medium for scientific exposition, diction, and enquiry, as in this account of the divisions of the sky:

> Besides a number great there is of sundry circles
> framed,
> That pass by both the foresaid *Poles, Meridians*
> rightly named,
> That over us directly runs; a Circle more doth
> lie
> The *Horizon* called: the world in midst divided
> is thereby . . .[2]

In addition to these translations, original works in English were produced by the likes of Robert Recorde, William Fulke, Leonard Digges, and Thomas Digges, which, as well as summarizing centuries' worth of scientific opinion, offered challenges to received wisdom and made reference to more recent developments. They will be a frequent source of reference later in this chapter.

While texts such as the ones named above have differences of emphasis and disagree about details, they broadly share a set of assumptions about the place of the Earth in the universe that derive from the works of the Greek philosopher Aristotle, albeit with significant modifications over the intervening centuries. These placed the Earth at the centre of the universe; the Moon, Mercury, Venus, the Sun, Mars, Jupiter, and Saturn revolving around it in concentric spheres; and the fixed stars circling together beyond them. Bodies above the Moon were unchanging and moved in regular circles, but sublunary things were subject to change and decay, made of the four elements fire, air, water, and earth, of which the first two naturally moved upwards, the latter two downwards. (For a sixteenth-century representation, see Figure 1.1.) The wide acceptance of this model reflects its congruence with everyday experience, according to which the Sun appears to rise in the east and travel across the sky to its setting, and when you drop something, it hits the ground.

Of course, everyday experience can be misleading: for example, although it feels stationary, the surface of the Earth is moving at 460 metres per second as the planet spins on its axis, simultaneously orbiting the Sun at over 100,000 kilometres per hour. However, the fundamentally inaccurate nature of their premises did not prevent ancient and medieval astronomers from calculating the movements of the heavens with great precision: the Oxford friar Nicholas of Lynn, for example, composed in 1386 'an astrological calendar valid for seventy-six years', including detailed astronomical data and predictions of eclipses.[3] In order to do this, they relied on techniques and tabulations developed by the Alexandrian Claudius Ptolemy around 150 CE and refined over the centuries that followed, by Muslim and Jewish astronomers in particular. Among other things, Ptolemy offered ways of dealing with aspects of heavenly motion that do not accord with Aristotle's regular, spherical model, such as the fact that planets sometimes seem to stop and then reverse their apparent movement around the Earth in the phenomenon known as retrogression. (The French King Charles alludes to this in *1 Henry VI* when he compares the capricious nature of the Roman god of war with that of his planetary namesake: 'Mars his true moving—even as in the heavens, | So in the earth—to this day is not known', 1.2.1–2.) As well as refining the existing concepts of epicycles (in effect, orbits within orbits) and

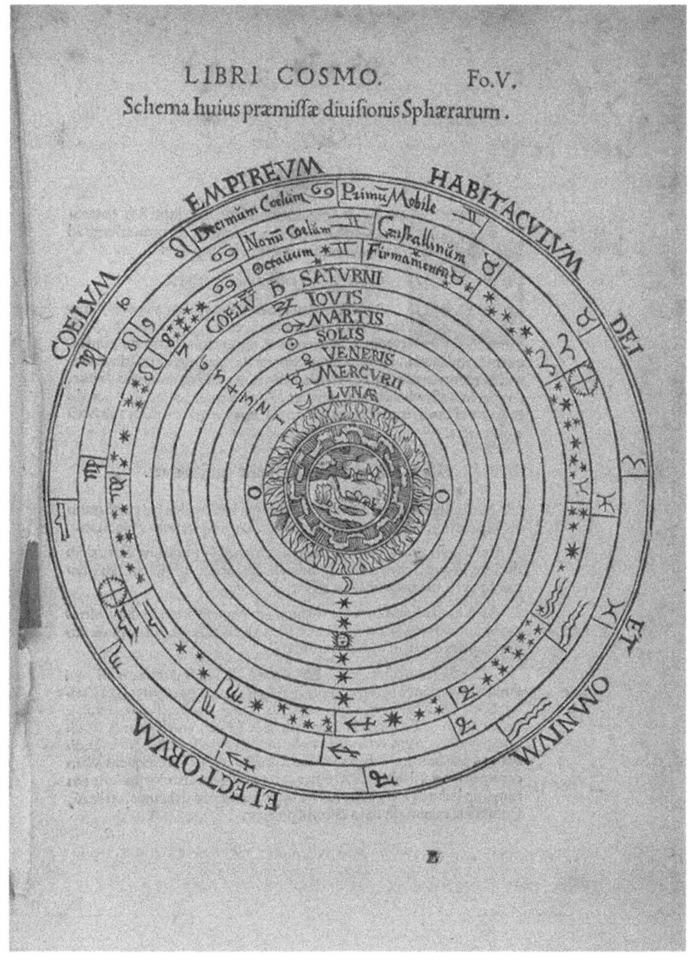

Figure 1.1. Diagram of the cosmos showing the earth at the centre of a sequence of spheres with that of Luna (the Moon) innermost. The outermost sphere or Primum Mobile completes its cycle in a single day, causing the diurnal movements of heavenly bodies. Peter Apian, *Cosmographia* (Antwerp, 1539). New York Public Library.

deferents (whereby a planet's motion centres round a point that is not identical with the position of the Earth), he explained the apparently irregular speed of a planet in its epicycle by relating it to a third point, the equant.[4] Although its complexity was at odds with the simple and regular system imagined by Aristotle, Ptolemy's model provided a reliable way of accurately calculating the movements of the heavens that kept the Earth at the (approximate) centre of things.

Another influential feature of Aristotle's cosmic model that turned out to be incorrect was the way he distinguished between the changeable things on and around the Earth, made of the four elements, and the unchanging nature of the heavens, which were understood to be made of a fifth element or quintessence. Meteorological events were explained as the result of the action of the Sun, Moon, planets, and stars upon earthly matter. In some instances these explanations are close to the ones accepted today, as with Aristotle's description of cloud formation in his *Meteorology*: 'The earth is at rest, and the moisture about it is evaporated by the sun's rays and the other heat from above and rises upwards: ... the vapour cools and condenses again as a result of the loss of heat and the height and turns from air into water: and having become water falls again onto the earth.' However, as well as describing the vapours that form through the action of heat upon moist parts of the Earth, Aristotle also makes reference to the hot, dry exhalations caused by the heating of dry places, which he uses to explain phenomena such as thunderbolts and shooting stars. His treatment of the latter as meteorological rather than astronomical reflects the assumption that nothing above the Moon is subject to change: a transitory phenomenon like a comet must therefore have been generated on Earth. Wind is also a dry exhalation, whereas earthquakes are caused by wind trapped beneath the earth, bringing them, too, within the province of meteorology.[5]

Shakespeare's works are often informed by Aristotelian assumptions about the Earth, its weather, and its place in the universe. A geocentric understanding of the cosmos underlies the lengthy simile Richard II uses to explain that following his return from Ireland, his rival Bolingbroke will be put to shame:

> know'st thou not
> That when the searching eye of heaven is hid
> Behind the globe, that lights the lower world,
> Then thieves and robbers range abroad unseen

> In murders and in outrage bloody here;
> But when from under this terrestrial ball
> He fires the proud tops of the eastern pines,
> And darts his light through every guilty hole,
> Then murders, treasons, and detested sins,
> The cloak of night being plucked from off their
>   backs,
> Stand bare and naked, trembling at themselves?
>   (3.2.32–42)

The grandeur of this description is heightened by its God's-eye view of the Earth, which is imagined not as the apparently flat surface on which we rest but—as if seen from space—as a 'terrestrial ball' orbited by the Sun. According to Aristotle, the position of the Earth at the centre of the universe is the inevitable consequence of the tendency of earth (the element) to gravitate to the centre: 'it is heavy and what is heavy remains at the centre, and the earth does so'.[6] Shakespeare often uses 'centre' in this cosmic sense, most frequently in *Troilus and Cressida* which, perhaps not coincidentally, is one of only two Shakespeare plays to refer to Aristotle by name (in this instance with notorious anachronism, at 2.2.165). Ulysses does so in his famous speech on degree in 1.3, and both lovers compare their mutual fidelity to earth's attraction to the centre: 'the strong base and building of my love | Is as the very centre of the earth, | Drawing all things to it', as Cressida puts it (4.2.29–31; see also 3.2.175). Furthermore, as S. K. Heninger contends, 'Shakespeare was quite knowledgeable in Aristotelian meteorology, and his works abound in the technical terminology associated with it.'[7] When 1 Henry IV asks the rebel Earl of Worcester to 'move in that obedient orb again | Where you did give a fair and natural light, | And be no more an exhaled meteor' (*1 Henry IV*, 5.1.17–19), he wants Worcester to behave like a regular heavenly body rather than a short-lived comet made of an exhalation drawn up from the earth. And the conceit in Longueville's sonnet in *Love's Labour's Lost* relies on the assumption that vapours, too, are 'exhaled' from the earth through the action of celestial bodies:

> Vows are but breath, and breath a vapour is.
> Then thou, fair sun, which on my earth dost
>   shine,

> Exhal'st this vapour-vow; in thee it is.
> If broken then, it is no fault of mine.
>   (4.3.65–68)

While Shakespeare did not attend university, where he could have read texts like Aristotle's *Physics*, *On the Heavens* or *Meteorology* (very likely in a Latin translation), he had ample opportunity to learn of his ideas in English works like those by Fulke and Recorde.[8] In addition, as will become apparent later in this chapter, astronomical and meteorological imagery derived from Aristotle was common currency in the Elizabethan theatre.

Challenges to Aristotle: Some Critical Approaches

In spite of the broadly Aristotelian terminology and imagery deployed in Shakespeare's works, a frequent and fruitful area of interest for critics working on his treatment of astronomy and meteorology has been the extent of his familiarity with the work of those in his time who rejected Aristotelian explanations. Scholars frequently position Shakespeare as living in a period 'when the settled … accounts of how the universe worked began to fall apart and the new ideas that would replace them were still in flux', as Mary Thomas Crane puts it in *Losing Touch with Nature: Literature and the New Science in Sixteenth-Century England*. For Crane, a 'watershed event' was the appearance of a new star (actually a supernova) in the constellation of Cassiopeia for a period of over a year starting in November 1572. 'Because the star was so large and bright, several astronomers were able to see that there was no observable parallax, proving that the new object must be located above the Moon in the firmament of fixed stars where change was supposed to be impossible.' Either the new star was a miracle, or the Aristotelian distinction between the sublunary and the superlunary did not hold. Crane reads work by writers including Shakespeare, Marlowe, and Spenser in light of these and other indications that 'lived experience of the world did not reliably provide access to the truth about it'.[9] A similar sense of Shakespeare as living in a time when 'The terrestrial globe itself was in the process of becoming unmoored' informs Kristen Poole's *Supernatural Environments in Shakespeare's England: Spaces of Demonism, Divinity and*

Drama. However, Poole is especially concerned with the implications of a growing 'understanding of space as mathematical and geometric' for religious questions such as the power of demons, the relationship between God and the universe, and the location of Purgatory (if it existed)—a topic that bears on the nature, reliability, and staging of the Ghost in *Hamlet*.[10] Critics have approached Shakespeare's treatment of the weather, too, with an eye on the changing meteorological knowledge of his day. An example is Steve Mentz, who notes how European colonizers' (or would-be colonizers') experience of unfamiliar types of storm 'changed ancient ideas about global weather. Through the gradual recognition that hurricanes were distinct from tempests, European meteorology became a global science.'[11] It was a hurricane that nearly sank the *Sea Venture* in 1609, generating the narrative that Shakespeare drew on for *The Tempest*, and while the word does not appear in that play it famously does in *King Lear* (and, no less anachronistically, in *Troilus and Cressida*). Elsewhere, Mentz focuses on the weather in *Lear* (a play that certainly invites such attention), contrasting its depiction with that in other Shakespeare plays as being radically 'strange': hostile, overwhelming, and resistant to human attempts (like Aristotle's) to reduce it to part of an orderly system.[12] Mentz's approach is avowedly ecocritical, and given the urgency of climate change as an environmental, political, and intellectual priority it is hardly surprising that other critics have discussed Shakespeare and weather in similar terms: examples are Gabriel Egan, who sees *Lear* and *The Tempest* as raising questions about human responsibility for the weather that gain new meanings in the context of the climate emergency, and Todd A. Borlik, who also notes historical factors informing Shakespeare's treatment of this theme, in particular the dire weather of 1594 and 1595 and its impact on harvests and food prices.[13]

An alternative critical approach to contextualizing Shakespeare's astronomy and meteorology in relation either to early modern or to current thought and experience is to consider the theatrical and literary effects Shakespeare achieves through his deployment of these bodies of knowledge. In *Shakespeare's Storms*, Gwilym Jones takes his cue from Leslie Thomson's contention that thunder and lightning effects in the playhouse typically serve to 'establish or confirm a specifically supernatural context in the minds of the audience'.[14]

Jones argues that this assumption is one that Shakespeare exploits not only to produce a supernatural ambience, but also to allow his characters to raise questions about the natural or supernatural causes of the weather.[15] As Cicero puts it in *Julius Caesar*, in response to Casca's report of 'a tempest dropping fire' and other events Casca takes to be 'portentous things', 'men may construe things after their fashion, | Clean from the purpose of the things themselves' (1.3.20, 31, 34–35). Sophie Chiari in *Shakespeare's Representation of Weather, Climate and Environment* similarly finds Shakespeare depicting humans trying to make sense of the weather and imposing their own meanings on it. However, weather is also something that imposes itself on humans and their decisions, as with the 'dog days' of summer during which *Romeo and Juliet* (unlike its sources) is set. The *dies caniculares*, so called because they coincided with the rising of Sirius in Canis Major, were associated not just with heat but with the choleric humours that prompted anger, as well as with plague and ill-health; they were also thought to be a dangerous time in which 'to use carnal copulation'. In terms of early modern science, the play's combination of sex and violence, as well as the outbreak of 'pestilence' that generates the tragic misunderstanding at its climax, all relate to the combined influence of the stars and the weather: 'climate acts as an agent of fate'.[16]

The critics named above often return to those Shakespeare plays that seem most naturally to invite analysis that focuses on astronomy and meteorology. These include *A Midsummer Night's Dream*, in which marital breakdown between Oberon and Titania precipitates bad weather; *Romeo and Juliet*, in which the action of fate on the 'star-crossed lovers' is often imagined in astrological terms; *Julius Caesar*, in which portentous atmospheric conditions are both staged and discussed; *King Lear*, where the storm embodies a universe that might be sympathetic, hostile, or simply indifferent to Lear's plight; and *The Tempest*, which depicts a magus who claims to control the weather. *Hamlet* has also seemed to some critics to reflect a contemporary uncertainty about the Earth's place in the cosmos and about the possibility of understanding it, and this chapter will accordingly culminate in a discussion of that play. First, however, it will focus on a group of plays that are less frequently linked to themes of astronomy and meteorology, namely Shakespeare's history plays of the 1590s.

These plays largely lack prominent atmospheric effects like the storms in *Julius Caesar*, *King Lear*, or *The Tempest*, although *3 Henry VI* does include a scene in which 'Three suns appear in the air' (2.1.20: stage direction from the 1595 octavo) and—as per Thomson's argument—thunder and lightning herald the entrances of devils in Part 1 (5.3) and Part 2 (1.4.23s.d.). However, events in the heavens make themselves felt in other ways, from the discussions of the appearance of the sun before the battles of Bosworth and Shrewsbury to the more homely moment in *1 Henry IV* when the First Carrier calculates the time of night by noting that 'Charles's Wain'—the group of stars also known as the Plough or Big Dipper—'is over the new chimney' (2.1.2). Celestial and meteorological imagery abounds in the Histories, partly because of the ubiquitous use of the Sun as a symbol for royalty, and also because of the way in which words that denote basic features of the cosmos or of weather—sun, rain, air—provide an abundant source of puns appropriate to plays' regal and dynastic concerns: son, reign, heir. However, the discussion that follows focuses on an aspect of the Histories that, it will be argued, fundamentally shapes its treatment of astronomy and meteorology, and that is their concern with Providence.

Providence, History, and Meteorology

In Christian theology, Providence refers to God's care for and guidance of the processes of the universe, from the movements of the heavens down to the smallest detail. As Jesus puts it in the Gospel of Matthew, in a passage echoed in the final scene of *Hamlet*: 'Are not two sparrows sold for a farthing, and one of them shall not fall on the ground without your Father? Yea, and all the hairs of your head are numbered' (Matthew 10:29–30, Geneva Bible). In practice, as Kristen Poole demonstrates, different theologians—even theologians on the same side of the confessional divide, like John Calvin and the Anglican Richard Hooker—could treat this concept with significant differences of emphasis.[17] While Hooker argues that God 'hath stinted th'effects of his power' in ordaining laws within which nature operates, Calvin criticizes those who 'restrain the Providence of God within so narrow bounds, as if he suffered all things to be carried with an ungoverned course according to a perpetual law of nature'.[18] Calvin emphatically denies that anything can be ascribed to nature or fortune; instead, God 'governing both heaven and earth, by his Providence so ordereth all things that nothing chanceth but by his advised purpose'. Although Calvin is understandably cautious in discussing the political consequences of this notion, they are unmistakeable: 'undoubtedly we ought to believe that whatsoever changes of things are seen in the world, they come by the secret stirring of the hand of God'.[19] By implication, military victories and defeats, the rise and fall of dynasties, and the births and deaths of monarchs all happen because God wants them to.

As numerous critics have recognized, the notion of Providence is of central importance to Shakespeare's Histories, concerned as they are both with questions of causality and with the human tendency to discern patterns in the unfolding of events.[20] But the presence of Providence as a theme also has implications for the plays' treatment of the weather, as Calvin indicates when he criticizes those who think that

> when sometime immoderate heats with dryness do burn up all the grain, sometime unseasonable rains do mar the corn, when sudden harm cometh by hail & tempests: it shall not be the work of God, unless parhap it be because the clouds or fair weather or cold or heat have their beginning of the meeting of the planets or other natural causes. But by this mean is there no room left, neither for the fatherly favour nor for the judgements of God.[21]

Identifying natural (or astrological) causes for the weather is misguided because it obscures its real origin in the will of God, who may use it to bless or punish those on the receiving end.

Although Calvin's description of a universe in which nothing happens by chance was highly influential, the diversity of opinion on the question of Providence (in particular, as it relates to the weather) is well illustrated in responses to the earthquake that struck England on 6 April 1580, causing some damage and loss of life. Arthur Golding, for example, whose translation of Ovid's *Metamorphoses* Shakespeare knew well, took issue with those who 'will not stick to deface the apparent working of God, by ascribing this miracle to some ordinary causes in nature'; rather, 'it must needs appear to be the very finger of God', admonishing sinners to repent, and threatening destruction to the unrepentant.[22] Similarly, Thomas Churchyard noted that 'some fine headed fellows will wrest (by natural arguments) Gods doing and works, to a worldly or earthly operation proceeding from a hidden cause in the body and bowels of the earth.... Yet those that fear God ... will take the Earthquake to be of another kind of Nature'.[23] Golding's rejection of 'ordinary causes', and Churchyard's insistence that the earthquake is 'of another kind of Nature' from the phenomena described by the Aristotelians, can seem inherently anti-scientific, positing a God who is always liable to disrupt the workings of nature in the interests of warning or punishment.

However, since the Middle Ages scholars had had the intellectual tools to combine a belief in Providence with an emphasis on the natural means by which phenomena were brought about. They achieved this by adapting Aristotle's thinking on causality, which distinguishes between the final, formal, material, and efficient causes of events: commentators could describe the action of heavenly bodies in generating the winds that caused earthquakes (for example) while simultaneously treating these as means by which God's will was achieved. As Thomas Twyne writes in his work on the earthquake,

following *Aristotle* as chief in this behalf: we must understand, that the efficient causes of an Earthquake are three, to wit, the Sun, the other six Planets, and a spirit or breath included within the bowels of the earth: and the material cause one, which is an Exhalation, that is to say, a certain air, breath, or smoke drawn out of the earth, which of nature is hot and dry.

Having identified these efficient and material causes, Twyne goes on to distinguish between them and the final cause, the will of God: 'he which is the cause of all causes, in all his works of nature hath made them his under Deputies, remaining nevertheless at his check, without any absolute authority of their own'.[24] A similar intellectual strategy is evident in a work that predates the earthquake, *A Goodly Gallery ... to Behold the Natural Causes of All Kind of Meteors*, by the theologian William Fulke. Fulke sets out detailed explanations of the natural processes behind meteorological phenomena, while identifying their 'final cause' as 'the universal chief and last end of all things', 'the glory of God'. Even the most outlandish weather events, while 'sent of God as wonderful signs, to declare his power, & move us to amendment of life', are 'indeed miraculous, but not yet so, that they want a natural cause'.[25] For Fulke and Twyne, God works through nature, not in violation of it.

The application of Aristotle's theory of causality was conducive to scientific enquiry in the Middle Ages and early modern period, in that it opened up a space in which natural causes could be discussed without violating orthodox religious premises. However, it also had implications that are neatly exemplified in Gabriel Harvey's discussion of the earthquake in *Three Proper, and Witty, Familiar Letters* between him and the poet Edmund Spenser published in 1580. In what is described as a 'short, but sharp, and learned Judgement of Earthquakes', Harvey, like Twyne, distinguishes between material, formal, efficient, and final causes, offering a conventional Aristotelian account of their underlying meteorology while also acknowledging that nature is ultimately controlled by God, who 'hath all these secondary inferior things, the four Elements ... ever pliable and flexible Instruments at his Commandment'. Harvey also agrees that the final cause of earthquakes may sometimes be 'to testify and denounce the secrete wrath, and indignation of God, or his sensible punishment upon notorious malefactors'. However, since God works through natural processes, it is impossible to be sure when an earthquake is a warning or punishment, and when it happens through the regular course of nature: 'I cannot see, and would gladly learn, how a man on Earth, should be of so great authority, and so familiar acquaintance with God in Heaven, ... as to be able in such specialties ... to reveal his incomprehensible mysteries'.[26] In contrast to Calvin's Providential universe in which nothing happens by chance, Harvey describes a universe in which the operation of God is impossible to identify with certainty.

As has already been noted, the concept of Providence was just as applicable to the understanding of political outcomes as to the interpretation of weather events. For example, in *The Union of the Two Noble and Illustrate Families of Lancaster and York*, a key source for Shakespeare's first tetralogy of Histories, Edward Hall attributes the death of Richard III to 'the very divine justice and providence of God, which called him to condign punishment for his scelerate merits and mischievous deserts'.[27] The analyses of historical causality that early modern Londoners were able to see in history plays were marked by similar debates about the possible role of God in human affairs to those that were evident in discussions of the 1580 earthquake. A pre-eminent example, and one by which Shakespeare was profoundly influenced, was Christopher Marlowe's two-part play *Tamburlaine the Great* (*c.* 1587–88), which depicts the rise of the Scythian shepherd-turned-warlord to kingship and, subsequently, imperial power. From early on, Tamburlaine insists that his victories—which occur in an uninterrupted sequence across the two plays—have been pre-ordained by God, who has appointed him the 'scourge' of reigning monarchs. This raises a number of questions with awkward implications for Elizabethan political and religious orthodoxy. Are there circumstances in which rulers (understood to be appointed by God) may be deposed? If the usurpation is successful, does that mean that the new monarch is favoured by God, who has used him as a means of bringing about His secret purposes? Or perhaps the whole idea that God intervenes in history like this is just an ideological tool that monarchs use to justify their position—as Tamburlaine, early in Part 1, uses the treasure and the Egyptian Sultan's daughter that he has just captured as signs sent by God to persuade Theridamas 'That I shall be the monarch of the East'.[28]

Marlowe frequently uses meteorological imagery as a means of exploring these questions. Not only does Tamburlaine remark of the supposedly God-given gold, 'See how he rains down heaps of gold in showers' (1.2.182); he repeatedly identifies himself and his army with the weather, as when he promises that 'our swords, our lances, and our shot' will 'Fill all the air with fiery meteors' (4.2.51–52). At the end of Part 1 he claims that the blood they have spilt in Africa has generated prodigious weather events, as 'swelling clouds, drawn from wide gasping wounds' have 'Been oft resolved in bloody purple showers' (5.1.459–460). And just as God was understood to use such events as warnings of future punishment, so the doomed Agydas realizes that Tamburlaine's eyes 'shine as comets, menacing revenge' (3.2.74). Images like these do not merely liken Tamburlaine to the weather, as a phenomenon sent by God to admonish humanity. They show Tamburlaine claiming a God-like ability to make the weather and using this claim to bolster his own authority.

Tamburlaine is highly unusual among plays of the late 1580s and early 1590s both for the scepticism with which it treats divine Providence and for the level of power it gives its protagonist, and other

dramas of the time bear its influence while proclaiming more orthodox attitudes to politics and religion. An example is George Peele's
Troublesome Reign of John, King of England, the prologue to whose
first part explicitly names Marlowe's play but which in the prologue to
Part 2 asserts the inexorable working of the heavens: 'Nor can earth's
pomp or policy prevent | The doom ordained in their secret will.' At
the end of the play the title character, poisoned by a monk, recognizes
his death as a 'plague | Inflicted on thee for thy grievous sins', which
include mistreatment of his subjects, plotting the murder of a rival to
the throne, and yielding to the authority of the Pope. The assumption
that God intervenes in human history informs characters' reactions in
Part 1 when 'five Moons appear' after John has had himself crowned
for a second time. John asks a man renowned for prophecy to interpret them, while two of the nobles who will revolt against him confer
more discreetly:

> PEMBROOKE   The heavens frown upon the sinful earth,
>   When with prodigious unaccustomed signs
>   They spot their superficies with such wonder.
> ESSEX   Before the ruins of *Jerusalem*,
>   Such Meteors were the Ensigns of his wrath ...

While for Peter, the Prophet of Pomfret, the moons signify different
countries' relationships with the Pope, for Pembroke and Essex, they
are signs of imminent destruction.[29]

As developed by writers like Marlowe and Peele, the history play
was a dramatic genre inherently interested in how the hand of God
might be identified in human affairs—whether in the working-out of
the plot as a whole, as the prologue to *The Troublesome Reign* suggests,
or, with whatever degree of insight, good faith, or political intent,
by individual characters. This made the staging and the discussion
of meteorological events, which could also be read as divinely produced, thematically appropriate. These were the generic norms that
Shakespeare inherited when working on the three parts of *Henry
VI*, *Richard III*, *King John*, and the second tetralogy of history
plays.

My discussion of Shakespeare's 1590s Histories will be approximately in order of composition, with one important exception: *1 Henry VI* is widely thought to have been written after Parts 2 and 3, versions of which first went into print as *The First part of the Contention Betwixt the Two Famous Houses of York and Lancaster* (1594) and *The True Tragedy of Richard Duke of York* (1595). Moreover, the extent of Shakespeare's contribution to Part 1 (in particular) has been widely questioned; in the recent *Authorship Companion* to *The New Oxford Shakespeare*, for example, Gary Taylor and Rory Loughnane suggest that it represents his revision of an earlier play by Marlowe and others.[30] This raises the question of whether *1 Henry VI* should be discussed in this chapter at all, and if so, where. I choose to begin with it, both because it makes the sequence of the first three plays (which I refer to as *Henry VI* Parts 1, 2, and 3) easier to follow, and because its likely status as a collaborative text by authors including Marlowe and Shakespeare (even if only as reviser) recognizes the extent to which Shakespeare's history plays draw on, and are enmeshed in, the works of other dramatists.

The First Tetralogy

1 Henry VI begins with the Duke of Bedford, brother of Henry V, lamenting the late King's death:

> Hung be the heavens with black! Yield, day, to
>   night!
> Comets, importing change of times and states,
> Brandish your crystal tresses in the sky,
> And with them scourge the bad revolting stars
> That have consented unto Henry's death—
> King Henry the Fifth, too famous to live long.
>   (1.1.1–6)

Bedford's call for the sky to darken in mourning is clearly hyperbolical rhetoric rather than a prediction of a likely meteorological event: compare the response to Tamburlaine's death uttered by his son Amyras,

'Meet heaven and earth, and here let all things end!'.[31] However, the lines that follow exemplify two widespread assumptions about the relationship between the heavens and the Earth: that comets could be portentous, 'importing change of times and states'; and that the positioning of the stars influenced human fortunes. In accusing the stars of 'revolting', Bedford says they have been disobedient in allowing Henry's death to happen, establishing a theme of rebellion that will persist throughout Shakespeare's history plays. But in astronomical terms, this is hard to imagine: if the stars are fixed in their respective positions within the sphere they inhabit, it is surely impossible for them to do anything other than turn with it. The word 'revolting' thus loses the sense of disobedience that it has in English and means something more neutral in line with its Middle French root *révolter*, to turn—raising the question of whether Henry's death is some kind of aberration or in line with the purposes of the creator of the universe.

Only a few lines after Bedford's eulogy, however, his uncle the Duke of Exeter criticizes his willingness to 'curse the planets of mishap' (1.1.23), and this shift of focus from the astronomical to the earthly prompts a squabble between the Duke of Gloucester and the Bishop of Winchester over the degree of the Church's support for Henry (in effect, whether the clergy prayed for him or preyed on him). While Bedford is beseeching Henry's ghost to 'Prosper this realm; keep it from civil broils; | Combat with adverse planets in the heavens' (1.1.53–54), a Messenger enters bringing news of the loss of French towns that had been held by England—reverses that the Messenger ascribes to 'No treachery, but want of men and money' (1.1.69), as well as factional disagreements among those running the country. The play's opening scene thus sets up two ways of reading the historical events it will depict: as a manifestation of the will of the heavens, and as a consequence of human factors such as greed, ambition, and incompetence.

A tension between these two readings persists throughout the play and is thematized most obviously in its treatment of Joan la Pucelle, who, announcing herself as 'the English scourge' (1.3.108), draws on the theatrical energies generated by Marlowe's Tamburlaine, and raises comparable questions. Her defeat of the French prince Charles in single combat is taken as proof of divine favour, and the English hero Talbot is astonished to see his troops put to flight by 'A woman clad in armour' (1.7.3). However, supernatural agency proves to be something that characters discern only when it suits them: accused of inconstancy when the English storm Orleans, Joan unabashedly asks whether 'Sleeping or waking must I still prevail' and lays the blame on the negligence of 'Improvident soldiers' (2.1.57, 59). Although the play literally demonizes Joan, eventually depicting onstage the devils with whom she is revealed to have been in league, this does not completely dispel the impression that both sides are capable of identifying supernatural intervention at their own convenience: Talbot is several times described as the 'scourge' of the French (2.3.14, 4.2.16, 4.7.77), who suspect him of being 'a fiend of hell' (2.1.47).

The somewhat ambivalent treatment of Joan in *1 Henry VI*—as a character who uses demonic assistance, but who also raises questions about how and why other characters identify the operation of divine will in human events—acquires an intertextual dimension during an episode in 3.2 when she is linked with meteorological phenomena. Along with four soldiers, she gains access to the city of Rouen disguised as peasants before 'thrusting out a torch from yonder tower' to show 'the best and safest passage in' for the French forces:

> *Enter Joan la Pucelle on the top, thrusting out a torch burning*
> JOAN   Behold, this is the happy wedding torch
>   That joineth Rouen unto her countrymen,
>   But burning fatal to the Talbonites.
> BASTARD   See, noble Charles, the beacon of our friend.
>   The burning torch in yonder turret stands.
> CHARLES   Now shine it like a comet of revenge,
>   A prophet to the fall of all our foes! (3.3.8–15)

For Charles, now King of France, the comparison of Joan's torch to a comet is merely a rhetorical figure: just as a comet might announce the fall of princes, so he hopes the torch that shines aloft will foretell English losses. For early playgoers, however, the image of a burning torch held out from an elevated position might have had more specifically theatrical associations. If *1 Henry VI* is the same play as the 'harey the vj' that premiered at the Rose playhouse on 3 March 1592, then it appeared at a time when Robert Greene and Thomas Lodge's play *A Looking Glass for London and England* was being revived at the same theatre: *A Looking Glass* was staged on 8 March, a day after *Harry's* second performance, and would be staged again on 27 March, 19 April, and 7 June before the theatres closed due to the plague, while *Harry* was regularly performed throughout this period, playing on 11 March, 16 March, 5 April, 13 April, and 21 April, plus a further six performances in May and June.[32] *A Looking Glass* is a biblical drama depicting the tyrannical behaviour of King Rasni of Nineveh, who persists in his misdeeds in spite of a series of what would appear to be unmistakeable divine warnings including a lightning bolt that kills his sister (whom he is in the process of marrying) and a flame of fire that consumes an arrogant courtier. Invariably, these portents are dismissed by Rasni's advisers as mere natural phenomena, as in this comment on the lightning bolt:

> These are but common exhalations,
> Drawn from the earth, in substance hot and dry,
> Or moist and thick, or Meteors combust,
> Matters and causes incident to time,
> Enkindled in the fiery region first.

In the context of Greene and Lodge's play, the explanations offered by the sages are clearly meant to be understood as misreadings: purely scientific accounts that, for reasons of sycophancy, ignore the self-evident status of these events as punishments or warnings. This is the case even when (in the words of a stage direction) 'A hand from out a cloud, threateneth a burning sword', a phenomenon they ascribe to 'clammy exhalations, | Or *retrograde* conjunctions of the stars'.[33] However, the characters' willed indifference to this phenomenon presumably did not prevent the audience from being impressed by a striking theatrical image, and perhaps being reminded of it when they saw Joan's hand holding a burning torch on the upper level of the stage. As a result, a stage gesture in *1 Henry VI* that is caused purely by human agency acquires overtones of a portentous, divinely created event—albeit with an outcome different from that hoped for by the French, who go on to lose Rouen.

The authors of *1 Henry VI* create a similar effect earlier on in the play in their staging of the deaths, by cannon shot, of the Earl of Salisbury and Sir Thomas Gargrave. The build-up to this incident is depicted in detail, as the Master Gunner of Orleans explains how the English are 'Wont through a secret grate of iron bars | In yonder tower to overpeer the city', and that he has spent the last three days waiting to see if they appear behind it. As his Boy takes over from him, Salisbury, Gargrave, and others appear 'on the turrets' and discuss the captivity and release of Lord Talbot before Salisbury invites the others to look down through the grate:

> Now it is supper-time in Orléans.
> Here, through this grate, I count each one,
> And view the Frenchmen how they fortify.
> Let us look in: the sight will much delight thee.
> (1.6.37–40)

As the English are discussing the 'best place to make our batt'ry next' (1.6.43), the Folio stage direction indicates 'Here they shoot, and Salisbury falls down.' Death by cannon shot is an unusual stage occurrence in early modern drama, and it seems likely that it produced an impressive theatrical spectacle. However, in its dreadful unexpectedness (for the victims), as well as its exploitation of playhouse pyrotechnics, it is also akin to effects such as the 'Lightning and thunder wherewith Remilia is strooken' in *A Looking Glass*, and it achieves the admonitory result that the lightning ought to have: 'O Lord have mercy on us, wretched sinners!' cries the mortally wounded Salisbury (1.6.48). The sense of divine punishment is compounded by the hubris with which Salisbury was looking down on the French a moment earlier, complacent in his supposed safety and motivated by the desire for 'delight' as much as military advantage. While the immediate cause of his death is unambiguously staged, the way it comes as a bolt from the blue has the implication that a God's-eye view like the one Salisbury wants to have is reserved for God alone. Indeed, a few moments later when Talbot is swearing vengeance there is 'an alarum, and it thunders and lightens' (1.6.75s.d.); his response, 'What stir is this? What tumult's in the heavens?' is followed by news that Charles and Joan la Pucelle, 'A holy prophetess new risen up', have come to raise the siege (1.6.75–76, 81). The noise and light in the heavens, which seem both to comment on Talbot's vow and to announce Joan's arrival, echo what we have just seen on Earth, blurring the distinction between natural phenomenon, message from God, and human military action.

At moments like these, the authors of *1 Henry VI* exploit theatrical technology to depict events that, while understood to be produced by human action, also look like divine interventions. Although these events are not, in themselves, meteorological phenomena, they exemplify a central concern of Shakespeare's Histories, namely the problem of how the hand of God can be distinguished in historical outcomes and in the meteorological occurrences that accompany them. They also exemplify the way in which Shakespeare's treatment of scientific themes was influenced not only by recognizably 'scientific' texts, but also by other kinds of writing: in this case, earlier history plays, with their particular concerns and conventions.

Notwithstanding the uncertain date and authorship of *1 Henry VI*, the questions it raises about the interpretation of sublunary events are ones that Shakespeare also takes up in other plays of the trilogy. Perhaps the most strikingly meteorological instance of this comes in Part 3, where the three suns that appear before Edward and Richard in Act 2, scene 1 prompt a discussion of what the spectacle might mean:

> EDWARD　　Dazzle mine eyes, or do I see three suns?
> RICHARD　　Three glorious suns, each one a perfect sun;
> 　　　　Not separated with the racking clouds,
> 　　　　But severed in a pale clear-shining sky.
> 　　　　See, see—they join, embrace, and seem to kiss,
> 　　　　As if they vowed some league inviolable.
> 　　　　Now are they but one lamp, one light, one sun.
> 　　　　In this the heaven figures some event. (25–32)

The appearance of multiple suns or parhelia is an actual phenomenon caused by sunlight being refracted through ice crystals in the Earth's atmosphere, and in dramatizing it Shakespeare drew on his source, Hall's *Chronicle*, which tells how on the day of the battle of Mortimer's Cross

the sun (as some write) appeared to the earl of March [i.e., Edward], like iii suns, and suddenly joined all together in one, and that upon the sight thereof, he took such courage, that he fiercely set on his enemies, & them shortly discomfited: for which cause, men imagined, that he gave the sun in his full brightness for his cognisance or badge.[34]

The possible causes of parhelia had been discussed by ancient writers including Aristotle and Pliny, and their theories were cited by sixteenth-century writers such as Thomas Hill and William Fulke, who wrote that the additional suns were 'Images of the sunne, represented in an equal, smooth and watery cloud, placed on the side of the sun'.[35] As well as asserting that multiple suns 'do naturally betoken tempest, and rain, to follow because they cannot appear, but in a watery disposition of the air', Fulke notes that 'For a supernatural signification, they have often times been noted to have portended the contention of Princes for kingdoms.'[36] For Edward, as for Richard, the supernatural signification outweighs any natural causes or effects: he reads it as inciting the sons of York to 'join our lights together | And over-shine the earth as this the world' (2.1.37–38), although Edward's decision to use three suns as his device is somewhat undermined by Richard's response, 'Nay, bear three daughters—by your leave I speak it— | You love the breeder better than the male' (2.1.41–42). Edward's interpretation is, furthermore, instantly invalidated by the arrival of a messenger announcing York's death, which means that the three suns cannot be understood to represent him, Edward and Richard. The irony here is Shakespeare's creation rather than deriving from Hall, where Edward has already heard of his father's death before the three suns appear. In reversing the sequence, Shakespeare generates a feeling of scepticism as to the possibility of making correct interpretations of prodigious events. That does not mean that the three suns are necessarily without significance: they would seem to exemplify convincingly (if rather belatedly) Fulke's association of them with 'the contention of Princes for kingdoms', while a historically minded playgoer might consider the three Yorkist monarchs (Edward IV, the short-lived Edward V, and Richard) who would eventually be supplanted by the Tudors. However, unlike the audience, none of the characters on stage is in a position to supply this reading. They assume that an unusual meteorological event means something, and perhaps it does; but Edward, at least, cannot help interpreting it in light of his own interests and predilections.

The scene with three suns takes into the realm of meteorology a theme that pervades the second and third parts of *Henry VI*, which is the question of how to interpret earthly phenomena. At one extreme is King Henry himself, whose characteristic response to notable events is to see the hand of God behind them and who is increasingly tormented by the suspicion that the civil strife racking his kingdom is a punishment for his grandfather's deposition of Richard II. To some extent this line of interpretation is undermined in the plays, such as when the apprentice Peter and his master, whom he has accused of saying the Duke of York is rightful king, are made to undergo trial by combat. Henry readily accepts the master's death as proof of his guilt—'God in justice hath revealed to us | The truth and innocence of this poor fellow' (2.3.106–107)—whereas York himself is more cynical, inviting Peter to 'thank God and the good wine in thy master's wame' (2.3.100–101). Henry's simple faith is set up in contrast to the percipience of his uncle, Humphrey, Duke of Gloucester, most obviously in their responses to a man who claims to have been cured of blindness at Saint Alban's shrine. While the King observes that 'God's goodness hath been great to thee' (as well as anxiously worrying lest 'by his sight his sin be multiplied', 2.1.86, 74), Duke Humphrey uses clever questioning to prove that Simpcox had been feigning blindness, before testing his alleged physical disability by having him whipped. The response of his enemy Cardinal Beauford to Simpcox's renewed vigour, 'Duke Humphrey has done a miracle today' (2.1.162), pays sarcastic tribute to Gloucester's ability to distinguish divine intervention from fraud; but it gains an edge from the sudden arrival of the Messenger bringing news that the Duke's wife has been arrested for 'Dealing with witches and with conjurors' (2.1.173). This reversal makes Gloucester's moment of triumph look a little hubristic, as if he is too confident in his powers of deduction—and, perhaps, too sceptical about the possibility of God intervening in human affairs. This is a possibility that the play never absolutely rejects: after all, the victory of Peter over his master may be treated comically, but (as Horner's confession proves) it is the right outcome.

The depiction of Duke Humphrey's use of forensic questioning against Simpcox makes it grimly appropriate that the most extensive attempt in the *Henry VI* plays to provide explanations through close observation of phenomena is the Earl of Warwick's description of his corpse, in which Warwick finds clear indications that Gloucester was murdered. Warwick explains how in death by natural causes the blood usually descends from the face to the heart;

> But see, his face is black and full of blood;
> His eyeballs further out than when he lived,
> Staring full ghastly like a strangled man;
> His hair upreared; his nostrils stretched with
>   struggling;
> His hands abroad displayed, as one that grasped
> And tugged for life and was by strength
>   subdued.
> Look on the sheets. His hair, you see, is
>   sticking;
> His well-proportioned beard made rough and
>   rugged,
> Like to the summer's corn by tempest lodged.
> It cannot be but he was murdered here.
> The least of all these signs were probable.
>   (3.2.168–178)

In its use of *enargeia*, or vivid descriptive rhetoric, Warwick's speech identifies a series of 'signs' suggesting that Gloucester was murdered. His reading of the corpse prioritizes establishing cause, in contrast to the more premonitory response of the King, who sees in it 'how deep my grave is made: | For with his soul fled all my worldly solace, | For seeing him I see my life in death' (3.2.150–152). However, while from a modern, scientific perspective Warwick might seem to represent the 'right' way of looking at Gloucester's corpse (perhaps making him the ancestor of the pathologist from TV crime dramas), the play does not privilege it above other possibilities. The definitive comment on his murder seems to come from the dying confession of the Cardinal, 'raving and staring as if he were mad' (3.3.0s.d.), which reworks features of Warwick's description: 'He hath no eyes! The dust hath blinded them. | Comb down his hair—look, look: it stands upright, | Like lime twigs set to catch my wingèd soul' (3.3.14–16). Warwick's deductions on the basis of bodily 'signs' give way to the inferences the King makes about the Cardinal's spiritual condition on the basis of his final delirium: 'Ah, what a sign it is of evil life | Where death's approach is seen so terrible' (3.3.5–6).

The shift into the final play of the first tetralogy brings with it a change of emphasis in the interpretation of earthly phenomena: deduction of the kind practised by Duke Humphrey and the Earl of Warwick falls by the wayside in a dramatic idiom that privileges the making and fulfilment of prophecies (notably those of Henry's widow Queen Margaret in the third scene) and warnings such as Stanley's dream in Act 3, scene 2. Or it might be more accurate to say that such deduction is forbidden by the operation of tyrannical power: as the Scrivener says of the indictment of Hastings that he was instructed to copy before Hastings had actually been arrested, 'Who is so gross | That cannot see this palpable device? | Yet who so bold but says he sees it not?' (3.6.10–12). Instead, Richard's future subjects are made to accept the validity of spurious physical evidence such as the late King Edward's 'lineaments, | Being nothing like the noble Duke my father' (3.5.89–90), which Richard uses to prove Edward's illegitimacy and exclude his son from the throne. The play sets up propagandistic manipulation on the one hand and, on the other, an obtrusive sense of divine justice working itself out that culminates in the sequence of ghosts of Richard's victims who appear on the eve of Bosworth telling him to despair and die.

The ghosts are, however, figments of Richard's dream rather than external supernatural manifestations: for all that the hand of Providence seems to be visible in the play's outcomes, these are achieved by the means of human agents. Lady Anne's injunction in the second scene, 'Either heav'n with lightning strike the murd'rer dead, | Or earth gape open wide and eat him quick' produces no *Looking Glass*-style intervention (1.2.64–65). The play does, nevertheless, offer one moment of ambiguously meaningful weather when, on the morning of Bosworth, Richard notes that the Sun has not risen when the 'calendar' says it should have: 'He should have braved the east an hour ago' (5.6.9) This produces in him a simultaneous sense of foreboding and an awareness that his response is irrational:

> The sun will not be seen today.
> The sky doth frown and lour upon our army.
> ... Not shine today—why, what is that to me

> More than to Richmond? For the selfsame
> heaven
> That frowns on me looks sadly upon him.
>   (5.6.12–13, 15–17)

Presumably Richard's analysis is correct: if the Sun has not risen yet, that is just as true for Richmond as it is for Richard, so it makes no sense to read it as a prediction of the outcome. Yet the fact that Richard, who up till now has been a manipulator of signs rather than an interpreter of them, cannot help thinking so indicates a great deal about his state of mind; and Shakespeare's shifts of focus between the two opposed sides leave it uncertain what actually is happening in the sky. In 5.4, Richmond utters conventional wisdom when he notes that 'The weary sun hath made a golden set, | And by the bright tract of his fiery car | Gives token of a goodly day tomorrow' (5.4.1–3): red sky at night, shepherd's delight. When, after the play has staged his and Richard's contrasting dreams, Richmond rises from his slumbers the following morning and asks the time, he is told it is 'Upon the stroke of four' (5.5.189), which could be twilight in the latter half of August but would be too early for the Sun to have risen above the horizon. This obscures the question of whether the weather is as he expected, so that the audience does not see him responding to the delay of the sunrise: while, rationally, he must experience it at some point, theatrically, it is something that only happens to Richard, later on in the same scene. Through sleight of hand, Shakespeare creates the impression of different, and meaningful, weather attending the two contenders to the throne.

The final scenes of *Richard III* show Shakespeare wanting to have it both ways: in so far as they end a decades-long cycle of violence allegedly shaped by a Providential logic designed to place Henry Richmond on the throne, they include a weather event that is made to look portentous, but at the same time, they offer a sceptical commentary on the whole notion that such an event might herald a victory for either side. The tension between these two interpretations is played out within the consciousness of Richard, who knows that the weather cannot be meaningful but is unable to stop himself thinking that it might be. This shift of focus away from the interpretation of meteorological phenomena to the psychology of human response to meteorological phenomena marks a change of direction that Shakespeare continues to pursue in *King John*.

King John, *Richard II*, and *Henry IV*

In dramatizing events from the reign of King John, Shakespeare took on materials that in his era came fraught with religious significance. Because he had defied the Pope over the installation of Stephen Langton as Archbishop of Canterbury, John could be seen as prefiguring the later, definitive break with Rome under Henry VIII, and his subsequent excommunication was a punishment that had been administered to the reigning monarch Elizabeth in 1570 and renewed in 1588. His eventual reconciliation with the Pope was read as a sign of moral frailty that was also manifested in bad government, a deficiency that John himself recognizes in Peele's *Troublesome Reign* when he tells himself, in a moment of prophetic insight:

> Thy sins are far too great to be the man
> T'abolish Pope, and Popery from thy Realm:
> But in thy Seat, if I may guess at all,
> A King shall reign that shall suppress them
>   all.[37]

The way Peele makes John look forward to the Henrician Reformation exemplifies the Providential interpretation of his reign encouraged by Protestant historiography.

Given such precedents, the ambivalence with which Shakespeare in *King John* treats the possibility of divine will being made manifest in observable phenomena is striking. As has already been noted, Peele dramatizes the five moons that according to Raphael Holinshed were 'seen in the province of York', offering an interpretation of them from Peter the Hermit (who in Holinshed does not appear for another twelve years) as signifying the Spanish, Danish, German, French, and English monarchies and their relationships with Rome.[38] Shakespeare, by contrast, keeps the moons offstage:

> HUBERT   My lord, they say five moons were seen tonight,
>   Four fixèd, and the fifth did whirl about
>   The other four in wondrous motion.
> KING JOHN   Five moons?

> HUBERT   Old men and beldams in the streets
>   Do prophesy upon it dangerously.
>   Young Arthur's death is common in their mouths,
>   And when they talk of him, they shake their heads,
>   And whisper one another in the ear;
>   And he that speaks doth grip the hearer's wrist,
>   Whilst he that hears makes fearful action,
>   With wrinkled brows, with nods, with rolling eyes.
>   (4.2.183–193)

This is a vivid description, but its liveliness consists in Hubert's images of human interaction rather than anything that is said about the celestial event itself. The moons, by contrast, are multiply framed: not only are they reported rather than shown, they are reported as hearsay ('they say'), and the consequent prophecies are implicitly diminished in being associated with 'Old men and beldams'. The description that follows is remarkable for its emphasis on manual workers, bodily confusion, and mental distraction:

> I saw a smith stand with his hammer, thus,
> The whilst his iron did on the anvil cool,
> With open mouth swallowing a tailor's news,
> Who, with his shears and measure in his hand,
> Standing on slippers which his nimble haste
> Had falsely thrust upon contrary feet,
> Told of a many thousand warlike French
> That were embattailèd and ranked in Kent.
>   (194–201)

A productive way of reading these lines, albeit one that derives from an analysis of a later historical period, comes from Steven Shapin's account of the development of evidentiary norms for the assessment of experimental data in the mid-seventeenth century. Shapin argues that the emerging discourse of early modern experimental science needed to find a way of convincing readers of the validity of experimental data, and it did so by relying on the testimony of educated males of high social standing.[39] The assumptions underlying that requirement are readily evident in Hubert's description, which devalues the speakers and hearers as labourers whose loss of bodily control implies an absence of critical faculty (as with the smith passively 'swallowing a tailor's news'). This casts doubt upon the stories they tell in a way that is not the case in *The Troublesome Reign*, where the five moons are seen by the King and senior nobles, as well as the audience.

The behaviour of the common people, as reported by Hubert, seems to bear out an earlier speech in which Cardinal Pandolf advises the French Prince Lewis of how he may capitalize on John's capture of his nephew (and rival claimant of the throne) Prince Arthur:

> John lays you plots; the times conspire with you
>
>   …
> This act, so vilely born, shall cool the hearts
> Of all his people, and freeze up their zeal,
> That none so small advantage shall step forth
> To check his reign but they will cherish it;
> No natural exhalation in the sky,
> No scope of nature, no distempered day,
> No common wind, no customèd event,
> But they will pluck away his natural cause,
> And call them meteors, prodigies, and signs,
> Abortives, presages, and tongues of heaven
> Plainly denouncing vengeance upon John.
>   (3.4.146, 149–159)

Pandolf's assumptions about popular credulity go further than the responses described by Hubert, in that they do not require prodigious events such as multiple moons as a stimulus: so volatile are the people that mere bad weather, a 'distempered day', or 'common wind', can be interpreted by them as a warning of God's vengeance. Shakespeare's treatment of meteorology here and in Hubert's description is different from what we see with the three suns in *Henry VI* (and indeed in the earlier plays' depiction of natural phenomena more generally) in that the explicit focus of interest is not the phenomenon itself, but the political question of how the common people might respond to it. This is something that can be exploited by a Pandolf or can create problems for a John.

Shakespeare continues to explore this topic in the second tetralogy of history plays, and its potential practical consequences are made apparent in *Richard II* when the Earl of Salisbury is attempting to levy troops in Wales for the absent king. As the Welsh Captain explains,

> 'Tis thought the King is dead. We will not stay.
> The bay trees in our country all are withered,
> And meteors fright the fixèd stars of heaven.
> The pale-faced moon looks bloody on the earth,
> And lean-looked prophets whisper fearful
>   change.
> … These signs forerun the death or fall of kings.
>   (2.4.7–11, 15)

The validity of this interpretation is one that Shakespeare treats with careful irony. The association of comets with changes of monarch was a commonplace: as Fulke writes, 'Generally it is noted of all Historiographers, that after the appearing of *Comets*, most commonly followed great & notable calamities. Beside this, they betoken (sayeth some) wars, seditions, changes of commonwealths and the death of Princes and noble men.'[40] In reading the comet as a sign that the King is dead, the Welsh prophets are wrong: he is still very much alive. However, in a broader sense their interpretation is correct, since Henry Bolingbroke will soon seize the throne and Richard will be dead within a few months. Or from another perspective, the comet could be seen as generating the events it seems to presage: it causes the Welsh to desert Richard, which causes Richard's surrender, which precipitates his loss of the crown. The idea that comets can generate self-fulfilling prophecies somewhat resembles Fulke's caveat about the belief that they herald calamities and changes:

For what times Comets do shine, there be many hot & dry Exhalations in the air, which in dry men kindle heat, whereby they are provoked to anger, of anger commeth brawling, of brawling fighting & war, of war victory, of victory change of commonwealths. Then also Princes, living more delicately than other men, are more subject to infection, therefore die sooner than other men.

Instead of seeing comets as signs of political change, he presents them as effects of the same causes that produce political change.

Fulke's emphasis on the natural causes of both political and meteorological phenomena parallels an important strand of the second tetralogy. A prominent example of this is the dialogue between Hotspur and Glyndŵr in *1 Henry IV*, Act 3, scene 1 that sees the irascible Hotspur take issue with Glyndŵr's account of the events that attended

his birth, when 'The front of heaven was full of fiery shapes' and
'The frame and huge foundation of the earth | Shaked like a coward'
(3.1.36, 15–16). Hotspur's first response is to deny any connection
between the two events: the same would have happened 'if your
mother's cat | Had but kittened, though you yourself had never been
born' (3.1.17–18). When Glyndŵr insists on the point, Hotspur
changes tack, offering a natural explanation:

> Diseasèd nature oftentimes breaks forth
> In strange eruptions; oft the teeming earth
> Is with a kind of colic pinched and vexed
> By the imprisoning of unruly wind
> Within her womb, which, for enlargement
>   striving
> Shakes the old beldam earth, and topples down
> Steeples and moss-grown towers. At your birth
> Our grandam earth, having this distemp'rature,
> In passion shook. (3.1.25–33)

While Hotspur's description of earth as a flatulent geriatric may seem
satirical, the substance of his explanation is in keeping with the opinion of Aristotle, who as well as ascribing earthquakes to subterranean
winds makes a similar comparison: 'the wind in the earth has effects
similar to those of the wind in our bodies whose force when it is pent
up inside us can cause tremors and throbbings, some earthquakes
being like a tremor, some like a throbbing'.[41] The clash between
natural and supernatural explanations recalls the questions raised in
responses to the earthquake of 1580.

To some extent Hotspur's deflation of Glyndŵr's rhetoric is
prompted by mere irritation: no scientific knowledge is required for
his subsequent rejoinder to Glyndŵr's claim to 'call sprits from the
vasty deep', 'But will they come when you do call for them?' (3.1.51,
53). However, his interest in the natural causes of earthquakes recalls
an earlier speech in which he complains that the Earls of Northumberland and Worcester, having helped Bolingbroke depose Richard
II, are now mistreated by him as King:

> shall it be
> That you a world of curses undergo,

> Being the agents or base second means,
> The cords, the ladder, or the hangman, rather?
> (1.3.161–164)

Hotspur's description of the earls as 'second means' employs the same language of causality that, as has already been noted, writers on meteorology used to refer to the natural processes by which God achieves his intended outcomes on Earth. In referring to them as Henry's 'agents', he effectively literalizes Thomas Twyne's metaphor, according to which 'the Sun, the Planets, and other Stars' are merely God's 'under Deputies … without any absolute authority of their own', thereby underscoring the exploitation of Northumberland and Worcester by the God-like Bolingbroke. Hotspur's familiarity with Aristotelian thinking on causality and on natural phenomena, as well as complicating any view of him as a 'mad fellow of the North' (2.5.338), exemplifies the second tetralogy's preoccupation with second causes, of political as well as of meteorological events.

Of course, Hotspur's reference to 'second means' when discussing Henry's rise to the throne is figurative, rather than literal: Hotspur doesn't really think Henry is God. But he may be implying that Henry does; or at least, that he identifies himself with divine purpose to an unhealthy extent. In *Richard II*, chastised by the Duke of York before Flint castle to 'Take not … further than you should, | Lest you mistake the heavens are over our heads', Henry responds, 'I know it, uncle, and oppose not myself | Against their will' (3.3.16.19). These lines, uttered as Hotspur appears on stage, are supremely ambiguous: they can be read as yet another insistence that Henry has returned from banishment simply to 'lay my claim | To my inheritance' (2.3.134–135), or as expressing the belief that God wants him to depose Richard. As the Bishop of Carlisle has already said to Richard, in a very different context, 'The means that heavens yield must be embraced, | And not neglected; else heaven would, | And we will not' (Additional Passage D, after 3.2.28). If Henry does secretly see himself as divinely appointed, that may explain the decidedly Tamburlaine-esque imagery he uses a few lines later when threatening, if his banishment is not repealed and lands restored, to 'lay the summer's dust with showers of blood | Rained from the wounds of slaughtered Englishmen' in what he describes as a 'crimson tempest'

(3.3.42–43, 45), recalling the 'bloody purple showers' Tamburlaine made in Africa. He continues:

> Methinks King Richard and myself should
>   meet
> With no less terror than the elements
> Of fire and water when their thund'ring shock
> At meeting tears the cloudy cheeks of heaven.
>   (3.3.53–56)

This is another imitation of Marlowe, reworking lines from the speech Tamburlaine makes when he ascends the Persian throne using the Turkish Emperor Bajazeth as his footstool:

> My sword struck fire from his coat of steel
> Even in Bithynia, when I took this Turk,
> As when a fiery exhalation,
> Wrapped in the bowels of a freezing cloud,
> Fighting for passage, makes the welkin crack,
> And casts a flash of lightning to the earth.[42]

In both cases, the encounter between the monarch and his nemesis is described as a weather event. Tamburlaine closely follows Aristotle's summary: 'Heat when radiated disperses into the upper region. But any of the dry exhalation that gets trapped when the air is in process of cooling is forcibly ejected as the clouds condense.'[43] Henry's description is similar in that lightning is the result of a hot exhalation meeting a damp cloud, although he reconceives this as a meeting of the elements of fire and water, which he identifies with Richard and himself. This allows him to generate the conceit, 'Be he the fire, I'll be the yielding water. | The rage be his, whilst on the earth I rain | My waters' (3.3.57–59), where the enjambment after 'rain' conceals an end-stopped line culminating in 'reign'—perhaps betraying Henry's true intent.

As well as revealing his ambition, Bolingbroke's meteorological simile distinguishes him from Richard in a similar way to how Hotspur's Aristotelian account of earthquakes distinguishes him from Glyndŵr. In both cases, it presents the characters as realists interested in the natural causes of things, as opposed to Glyndŵr with his mystical self-importance or Richard's belief that 'The breath of worldly men cannot depose | The deputy elected by the Lord'

(3.2.52–53). It would be an overstatement to regard the surrounding plays as uncritically endorsing this attitude: for instance, the unexplained music that Glyndŵr is able to summon seems to point to the existence of forces Hotspur does not understand. But it is fair to say that, especially from *King John* onward, Shakespeare offers accounts of the natural causes of meteorological events in a way that sometimes parallels and sometimes intersects with his interest in the causes of historical change. Individual characters might allude to the workings of Providence, as when York submits to Bolingbroke's authority on the grounds that 'heaven hath a hand in these events, | To whose high will we bound our calm contents' (5.2.37–38). But the plays themselves mirror Fulke's approach in *The Goodly Gallery* of treating final causes as ultimately unknowable and focusing on the more immediate causes of historical and weather events alike.

Hamlet

In his history plays of the 1590s, the natural explanations that are offered for meteorological phenomena are broadly in keeping with Aristotelian principles, as with Hotspur's account of earthquakes. *Hamlet*, however, is a play that critics often see as caught between traditional and emergent views of the cosmos. For Tom MacFaul, Hamlet's admonition after seeing the Ghost, 'There are more things in heaven and earth, Horatio, | Than are dreamt of in our [Quarto 'your'] philosophy' reflects the characters' sense of 'wonder as their system of knowledge is broken', while for Kristen Poole, his description of man as a 'quintessence of dust' (2.2.310), in collapsing the distinction between earthly materials and the celestial fifth element, 'crashes a familiar Aristotelian cosmology'.[44] Howard Marchitello similarly positions the play in relation to contemporary epistemological crisis, namely an 'increasingly serious skepticism over the viability of the body and its perceptions as the mechanisms that secure knowledge' in the late sixteenth century. Machines like those in Tycho Brahe's observatory on the island of Hven (a few miles south of *Hamlet*'s Helsingør, in the Øresund strait) potentially offer a more reliable means of understanding the universe, and Hamlet's deployment of the 'Mousetrap' to test the reliability of the Ghost looks forward to the attitudes of experimental science.[45] Carla Mazzio, however, sees the 'Mousetrap' as an inadequate experiment, 'designed to produce the effects it desired'.[46] She finds in the play's preoccupation with air, an element particularly resistant to observation, manipulation, and representation, a pre-eminent example of its pessimism about understanding and acting upon the natural world.

Mazzio is interested primarily in the 'meteorological debates of the period', and indeed the connection she makes in passing between references to Purgatory in *Hamlet* and in Fulke's *Goodly Gallery* would repay further research.[47] However, the nocturnal scene with which the play begins focuses attention, rather, on the astronomical: Marcellus has invited Horatio

> With us to watch the minutes of this night,
> That if again this apparition come
> He may approve our eyes and speak to it.
>   (1.1.25–27)

The word 'minutes' here is ambiguous. It can refer to the measurement of time, in which a minute is the sixtieth part of an hour; or of space, in which it is the sixtieth part of a degree. As time passes, the stars move by degrees as they appear to rotate in their heavenly sphere, until—

> BARNARDO Last night of all,
>   When yon same star that's westward from the
>   pole
>   Had made his course t'illume that part of
>   heaven
>   Where now it burns, Marcellus and myself,
>   The bell then beating one—
>   *Enter the Ghost.* (1.1.33–37)

The Ghost appears on its astronomical cue like the automated figure on a clocktower, just as it will disappear 'on the crowing of the cock' (1.1.138). The opening of the play foregrounds the observation of the heavens that leads up to its dramatic entrance, creating an implicit parallel between the celestial globe whose apparent turnings are measured by Barnardo, Marcellus, and Horatio, and the Globe theatre in which the night-time spectacle is imagined through the agreed complicity of actors and audience. As Poole puts it, 'The Globe was ... a container for examining the container of the cosmos'.[48]

The viability of Poole's analogy depends on an Aristotelian cosmology that places the Earth in the midst of a series of concentric spheres, with the fixed stars turning in their sphere beyond the planets. As has already been noted, Shakespeare's plays overwhelmingly accept this model of the universe, a tendency that is exemplified in *Hamlet* when Polonius vows to 'find | Where truth is hid, though it were hid indeed | Within the centre' (2.2.160–161). When Claudius tells Hamlet that his 'intent | In going back to school in Wittenberg, | … is most retrograde to our desire' (1.2.112–114), he similarly employs a Ptolemaic model, which, as has already been noted, used the idea of retrogression to explain planetary movements that did not fit in with geocentric assumptions. However, as Crane points out, 'ability to imagine the physical structure of the universe in Copernican terms is not, in fact, a reliable criterion for determining whether or not the "new astronomy" influenced the person doing the imagining'.[49] There are some hints in *Hamlet* that its playwright was familiar with the debate over heliocentrism, and expected his audience to be. For one thing, while the fact that Hamlet is a student at Wittenberg is undeniably significant for the fact that this was also the university of Martin Luther, Claudius's use of the astronomical metaphor of retrogression in relation to Hamlet's desire to return there also seems to point to its association with Copernican astronomy.[50] A more doubtful allusion is in the names of Rosencrantz and Guildenstern, which they share with Danish noblemen whose crests appear on the title page of Tycho's *Astronomiæ instauratæ mechanica* (1598). Finally, there is the poem Hamlet sends Ophelia:

> Doubt thou the stars are fire,
> Doubt that the sun doth move,
> Doubt truth to be a liar,
> But never doubt I love. (2.2.116–119)

As Marchitello points out, the verb 'doubt' here works in multiple ways that are enabled by its contradictory early modern meanings, 'To be in doubt or uncertainty' (*OED* 'doubt', *v.*, 1) and 'To dread, fear, be afraid of' (*OED* 'doubt', *v.*, 5).[51] The genre of love poetry implies sense 1 for the final line, that is, 'Never be in doubt as to whether I love', and the logic of this when read in combination with the second line seems to assume a geocentric universe: 'You may doubt even something obvious like the movement of the Sun, but never be in doubt as to whether I love.' However, in order for the third line to work, sense 5 has to be employed: 'You may fear something impossible, like truth being a liar, but never be in doubt as to whether I love.' The opposing meaning that is brought into play here contaminates the other meanings of 'doubt' in the poem, generating new possibilities: that truth is a liar, that the poem's speaker does not love, that the Sun does not, in fact, move.

Slight though this quatrain is, it is potentially productive as a basis on which to examine *Hamlet*'s treatment of astronomy because it emphasizes, not certainty about the Copernican system, but doubt as to the whole question of whether the Sun moves or not. When one examines the English-language texts on this subject to which Shakespeare could conceivably have had access, a recurrent theme is how hard it is to understand the evidence for heliocentrism if one lacks the mathematical training. For instance, the brief discussion of Copernicus in Robert Recorde's *The Castle of Knowledge* (1556)—which comes several hundred pages into a dialogue between a Master and a Scholar who has already assimilated some fairly complex information about the structure of the heavens, their different appearances at different latitudes, and how to make a celestial sphere—goes like this:

> MASTER　Copernicus, a man of great learning, of
> much experience, and of wonderful diligence in
> observation, hath renewed the opinion of
> Aristarchus Samius, and affirmeth that the earth not
> only moveth circularly about his own centre, but also
> may be, yea and is, continually out of the precise
> centre of the world 38 hundred thousand miles: but
> because the understanding of that controversy
> dependeth of profounder knowledge then in this
> Introduction may be uttered conveniently, I will let it
> pass till some other time.
>
> SCHOLAR　Nay sir in good faith, I desire not to hear
> such vain phantasies, so far against common reason,
> and repugnant to the consent of all the learned
> multitude of Writers, and therefore let it pass for
> ever, and a day longer.
>
> MASTER　You are too young to be a good judge in so
> great a matter: it passeth far your learning, and theirs

> also that are much better learned than you, to
> improve his supposition by good arguments, and
> therefore you were best to condemn nothing that you
> do not well understand: but another time, as I said, I
> will so declare his supposition, that you shall not only
> wonder to hear it, but also peradventure be as earnest
> then to credit it, as you are now to condemn it.[52]

Recorde's text, which has been described as 'the outstanding introduction to the science of astronomy published during the sixteenth century', declines to offer the evidence for heliocentrism on the grounds that it is way beyond the abilities of the junior speaker (and, by implication, the reader).[53] A similar awareness that the average reader simply may not have the maths to deal with Copernicus is on display in Thomas Digges's 'perfit description of the Celestial Orbs' (see Figure 1.2), appended to his father Leonard's *A Prognostication Everlasting* in editions from 1576 onwards. After summarizing Copernicus's arguments, Digges asserts that 'all these things, although they seem hard, strange, & incredible, yet to any reasonable man that hath his understanding ripened with Mathematical demonstration, *Copernicus* in his *Revolutions* according to his promise hath made them more evident and clear than the Sun beams'. However, he goes on to translate a section of Copernicus that offers 'reasons philosophical'— as opposed to mathematical—'alleged for the earth's stability, and their solutions, that such as are not able with *Geometrical* eyes to behold the secret perfection of *Copernicus' Theoric*, may yet by these familiar natural reasons be induced to search farther'.[54] Like Recorde, Digges is aware that the mathematical argument for heliocentrism may be beyond many readers' capacity.

It is quite possible that Shakespeare was familiar with Digges's description. Although Thomas Digges died in 1595 his widow Anne went on to marry Thomas Russell, whom Shakespeare appointed as an overseer of his will; Digges's son Leonard was an admirer of Shakespeare and composed commendatory verses for the First Folio in 1623. There are, then, tangible connections between Shakespeare and the Diggeses. However, perhaps more important than the question of direct knowledge is the reminder Digges's text gives of the difficulty most readers would have had in understanding Copernicus's evidence. This is also apparent when Copernicus is mentioned in another text

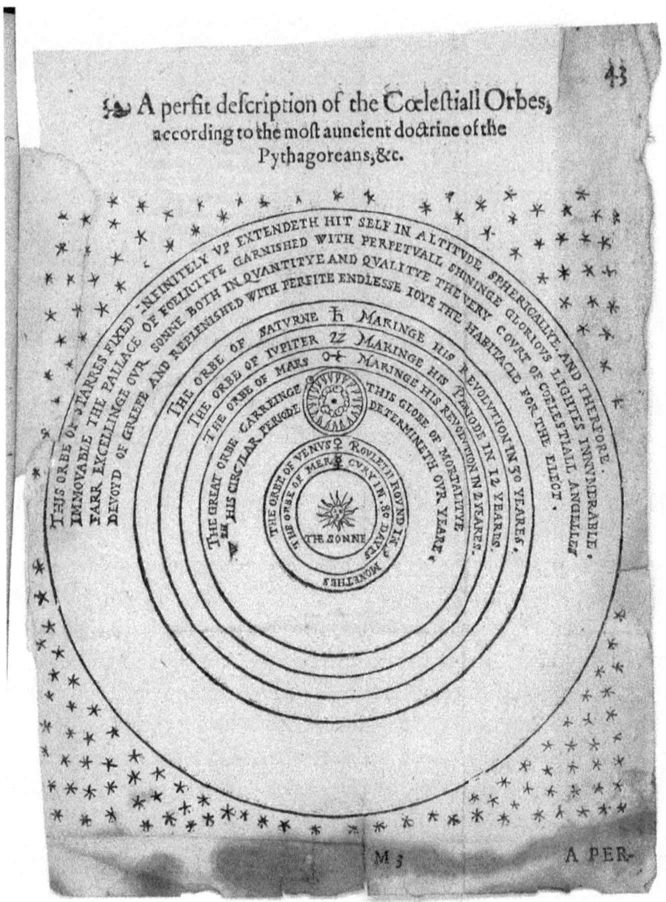

Figure 1.2. Diagram of the cosmos showing the sun at the centre. Thomas Digges, 'A perfit description of the Caelestial Orbs', in Leonard Digges, *A Prognostication Everlasting* (London, 1605). Wellcome Collection.

that Shakespeare demonstrably knew at least by the end of his career, the *Essays* of Montaigne:

The heavens and the planets, have moved these three thousand years, and all the world hath taught us so, until *Cleanthes* the *Samian*, or else (according to

Theophrastus) *Nicetas* the *Syracusian*, took upon him to maintain, it was the earth that moved, by the oblique circle of the *Zodiac*, turning about her axletree. And in our days *Copernicus* hath so well grounded this doctrine, that he doth very orderly fit it to all Astrological consequences. What shall we reap by it, but only that we need not care, which of the two it be? And who knoweth whether a thousand years hence a third opinion will rise, which happily shall overthrow these two precedent.[55]

Montaigne's remarks are invaluable firstly, in showing that a sixteenth-century author could reasonably expect his readers to be familiar with the arguments of Copernicus, but secondly, in their treatment of those arguments as no more than an 'opinion' that later discoveries might 'overthrow'. His response gives a notion of what a reader lacking Digges's '*Geometrical* eyes' might make of Copernicus's model.

John Florio's translation of Montaigne was unpublished at the time when Shakespeare was writing *Hamlet*, although it is possible that he saw the work in manuscript. However, it offers a useful way of approaching the play in that it shows how Copernicanism might prompt not a wholesale rethinking of one's cosmological assumptions but rather a resigned scepticism about the possibility of knowing the truth. Hamlet's comment on one of the skulls thrown up by the Gravedigger, while not intended as a verdict on heliocentrism, seems very apposite: 'Here's fine revolution, an we had the trick to see't' (5.1.88–89). Like the degeneration of 'my lord such a one' to undignified matter (5.1.83), the revolution of the Earth is something that, from his terrestrial vantage point, Hamlet does not have the 'trick' to observe. However, an awareness of it as a possibility may underlie the sense of decentredness to which Hamlet frequently gives expression, perhaps most fully in his words to Rosencrantz and Guildenstern:

I have of late—but wherefore I know not—lost all my mirth, forgone all custom of exercise; and indeed it goes so heavily with my disposition that this goodly frame, the earth, seems to me a sterile promontory. This most excellent canopy the air, look you, this brave o'erhanging, this majestical roof fretted with golden fire—why, it appears no other thing to me than a foul and pestilent congregation of vapours. (2.2.297–305)

The conception of the Earth as a 'goodly frame' constructed by the divine craftsman was a commonplace by the time of *Hamlet*; Recorde, citing Aristotle, uses a similar image to describe the universe when he refers to it as 'an apt frame of heaven and earth'.[56] However, for Hamlet this frame occupies a position that is not central, but marginal: a 'promontory' or outcrop, just as the air, rather than being designed as a 'roof' to 'canopy' it, is both unpleasant and disorderly. A similar feeling of decentredness can be identified in Hamlet's complaint, 'The time is out of joint. O cursèd spite | That ever I was born to set it right!' (1.5.189–190): while the metaphor can be understood in medical terms, the sense of a mechanism whose parts are out of order is supported by Claudius's use of the same image when he says that Fortinbras thinks 'Our state to be disjoint and out of frame' (1.2.20). The Copernican model offers Hamlet spatial metaphors with which to express his feeling of existential disquiet and political opposition. And given his playing on the homophones 'sun' and 'son' at 1.2.67 and 2.2.186, it may not be entirely outlandish to see in his feeling that (as John Donne famously put it) 'The sun is lost, and th' earth, and no man's wit | Can well direct him where to look for it' a feeling of resentment that, despite being the son of a king whom he twice compares to the sun-god Hyperion, he is not in the position of royal authority that he deserves.[57]

In its treatment of celestial phenomena, *Hamlet* offers a striking contrast to Shakespeare's history plays of the 1590s. The earlier texts are especially concerned with causes: with the variously natural and supernatural explanations that humans supply for the things they see in the sky (and, in the case of earthquakes, feel beneath the earth). These different types of explanation parallel the plays' interest in whether divine or human agents shape the pattern of history, and the possibility of opening out a space in which to analyse scientific and historical phenomena alike without reference to the will of God. *Hamlet*, however, problematizes not simply the causes, but the phenomena themselves, exploiting an uncertainty about the arrangement of the universe itself so as to provide a counterpoint to the Prince's sense of displacement and unease. Rather than offering a new vision of the universe grounded in observation, advances in astronomy provoke melancholy, scepticism, and doubt. As the next chapter will show, it was not only in the field of astronomy that profoundly different accounts of nature of things were available in Shakespeare's time: such uncertainty extended beyond the configuration of the universe to include the basic question of what it was made of.

| 2

Matter

> If the dull substance of my flesh were thought,
> Injurious distance should not stop my way;
> For then, despite of space, I would be brought
> From limits far remote where thou dost stay.
> No matter then although my foot did stand
> Upon the farthest earth removed from thee;
> For nimble thought can jump both sea and land
> As soon as think the place where he would be.
> But ah, thought kills me that I am not thought,
> To leap large lengths of miles when thou art gone,
> But that, so much of earth and water wrought,
> I must attend time's leisure with my moan,
> Receiving naught by elements so slow
> But heavy tears, badges of either's woe.

The speaker of Sonnet 44, frustrated at the fact of being 'far remote' from the person he wants to be with, fantasizes about being immaterial. If he were unconstrained by the physicality of his body's 'dull substance', he could instantly 'jump both sea and land' to arrive 'where thou dost stay' simply by thinking himself there, even if positioned on the far side of the world. The miraculous movement through space that the speaker imagines is partially enacted in the poem, which addresses the beloved directly and in the present tense: the thoughts of this speaker are evidently with the person he loves. But as the speaker laments in line 9, 'I am not thought': wherever his thoughts might be, other aspects of his selfhood, notably 'the dull

Shakespeare and Science. Tom Rutter, Oxford University Press. © Tom Rutter (2024).
DOI: 10.1093/oso/9780192898548.003.0003

substance of my flesh' and the 'I' who is somehow detachable from the thoughts he is thinking, are constrained by 'injurious distance', and the immediacy of 'thou dost stay' in line 4 has given way by line 10 to the recognition that 'thou art gone'. Rather than being made of pure thought, the speaker is 'of earth and water wrought', elements whose sluggishness and materiality limit him to moaning and tears.

In its assumption that material things, including the speaker of the poem, are made of the 'elements'—fire, air, water, and earth— Sonnet 44 is in accordance with the model of the universe referred to in Chapter 1 and which I have associated (very broadly) with Aristotle. In *On Coming-to-be and Passing Away* (often referred to as *On Generation and Corruption*), Aristotle argues that the 'matter of which … perceptible bodies consist … is not separable but always accompanied by contrariety, and it is from this that the so-called elements come into being'. The primary contrary qualities that exist in matter are heat, cold, dryness, and moisture, and these 'have attached themselves to the apparently simple bodies, Fire, Air, Water and Earth', fire being hot and dry; air, hot and moist; water, cold and moist; and earth, cold and dry. Of these, 'Fire and Air form the body which is carried along towards the "limit" [i.e., upwards], while Earth and Water form the body which is carried along towards the centre.'[1] Aristotle's identification of these four elements follows the pre-Socratic philosopher Empedocles, although Aristotle notes that Empedocles placed fire in opposition to the other three.

Shakespeare did not need to have read Aristotle in order to conceive of nature as made up of the four elements: the conception was widespread and conventional, appearing, for example, in a text he demonstrably did read, Ovid's *Metamorphoses* (here given in Arthur Golding's 1560s translation):

> This endless world contains therein, I say,
> Four substances of which all things are gendered. Of these four
> The earth and water for their mass and weight are sunken lower;
> The other couple, air and fire, the purer of the twain,
> Mount up, and nought can keep them down.[2]

The distinction between heavy earth and water and the 'purer' air and fire that 'Mount up' informs Sonnet 45, which takes up from where 44 left off: the latter two elements 'Are both with thee wherever I abide; | The first my thought, the other my desire'. The speaker of these two sonnets thus uses the elemental theory of matter as a means of exploring his conception of a selfhood that is fundamentally split. Like 'all things', he is made up of the four elements (compare *Twelfth Night*, 2.3.9: 'Does not our lives consist of the four elements?'). However, according to the conceit of Sonnet 45, those elements are not all in the same place, since the two that make up his airy thoughts and his fiery desire are not with him but with his beloved, threatening fatal consequences: 'My life, being made of four, with two alone | Sinks down to death, oppressed with melancholy.' This notion of a body being split between two locations is just as impossible as the transmutation of flesh into thought that the speaker envisaged in Sonnet 44; both sonnets pick up elemental theory only to reject its underlying principles in the interests of generating striking and hyperbolical images of internal division. This subordination of theoretical consistency to witty effects is in keeping with the pun in line 5 of Sonnet 44, which brings together the irrelevance of spatial separation if the speaker's fantasy were to come true—'No matter then'—and the dematerialization that is imagined as facilitating it: he would like to be, literally, 'No matter'. The phrase might also be read as reflecting the insouciance with which these two poems treat elemental theory. If creative misreadings generate good poetry, does it matter if they ignore fundamental details of the body of thought that informs them?

That question recalls the discussion in the Introduction to this book of whether it is appropriate to think of Shakespeare's writing as 'scientific' at all. Not only does it rely on models of the universe that are based on conjecture rather than experiment and that have now been disproved; it is not even consistent in its use of those models. For example, as will become apparent in the first part of this chapter, Shakespeare's sonnets draw on a number of different ways of thinking about matter: not just elemental theory but Ovidian images of matter as constantly changing, ideas drawn from alchemy, and debates about the Eucharist. Rather than being underpinned by a consistent and fully thought-out model of the universe as in the great works of Dante or Milton, they make use of an eclectic range of sources that they happily reshape and discard. Furthermore, some of those sources seem to belong to the field of poetry or religion, and this raises additional questions about whether texts that allude to them are really dealing with 'science' in any recognizable sense.

However, as this chapter will explain, the arguments underlying medieval conceptions of the Eucharist were articulated in surprisingly detailed terms derived from Aristotelian theories of matter. Furthermore, as has already been noted in the Introduction, the distinction between 'literature' and 'science' was not so obvious in the early modern period as it has since become. This fact is readily exemplified in a work that will be discussed in the second section of the current chapter, namely the *De rerum natura* of the Roman poet Lucretius. Written in the first century BCE and rediscovered in 1417, this text set out the atomist theories of the Greek philosopher Epicurus and would influence seventeenth-century scientists including Pierre Gassendi and Isaac Newton. Shakespeare's familiarity with *De rerum natura* has been a source of regular speculation over recent years, and the current chapter surveys the debate in the context of a discussion of Shakespearean tragedy.

The final part of the chapter moves on to discuss a play that brings the elements to the fore in juxtaposing an airy spirit with an abused slave who is addressed as 'earth' (1.2.316). The master of these characters, *The Tempest*'s Prospero, defies straightforward definition, identifiable both as a Faustian magician and as a Baconian scientist, while the power he wields can be read as a form of colonial domination and as theatrical illusion (to give only a few examples). He thus testifies to the fragile boundary separating early modern science from magic, as well as the extent to which scientific ambitions were embedded in political ones. The insistent metatheatricality of the play in which he appears implies a suspicion of science's claims to be able to act on the physical world, but also underlies the aptness of theatre as a means of exploring the relationship between the material and the immaterial.

The Sonnets

Shakespeare's use of elemental theory in Sonnets 44 and 45, including the distinction between earth and water (which move downwards towards the world's centre) and fire and air (which move upwards), is consistent with his practice elsewhere in his writing. Cleopatra, for example, prefaces her suicide by announcing:

> Husband, I come.
> Now to that name my courage prove my title.
> I am fire and air; my other elements
> I give to baser life. (5.2.282–285)

She strives to turn defeat into victory by treating the act of suicide as elevated and dignified, resolving at 4.16.88–89 'what's brave, what's noble, | Let's do it after the high Roman fashion'. This makes it appropriate for her to identify the aspect of herself that wills the deed, and which she sees as joining Antony after death, with the upward-tending elements, and to imagine the body that she leaves behind as a compound solely of earth and water. A similar division is made in a more humorous context by the Dauphin in *Henry V* during his extravagant praise of his horse:

When I bestride him, I soar, I am a hawk; he trots the air…. It is a beast for Perseus. He is pure air and fire, and the dull elements of earth and water never appear in him, but only in patient stillness while his rider mounts him. (3.7.15–16, 20–23)

The Dauphin praises his horse's fleetness of foot both by comparing it to Perseus's winged horse Pegasus and by insisting that its body consists only of the purer elements of air and fire. Earth and water are present in it not substantially but only figuratively in its good behaviour when mounted.

In the Sonnets, the explicit engagement with the elemental theory of matter that is evident in 44 and 45 is only one instance of a wider preoccupation with substance and materiality. In the first seventeen sonnets, for example, the speaker urges procreation as a means of overcoming a decline and death that are imagined in emphatically physical terms: 'thou art much too fair | To be death's conquest and make worms thine heir' (Sonnet 6). Later, when the emphasis shifts away from sexual reproduction, the speaker repeatedly meditates on the 'substance' of the fair youth, as in 37 and 53. And, as he asserts the

power of verse to confer immortality, he depicts a world where things
that seem solid are continually subject to change and decay:

> I have seen the hungry ocean gain
> Advantage on the kingdom of the shore,
> And the firm soil win of the wat'ry main,
> Increasing store with loss and loss with store
>   ... (Sonnet 64)

> O how shall summer's honey breath hold out
> Against the wrathful siege of battering days
> When rocks impregnable are not so stout,
> Nor gates of steel so strong, but time decays?
>   (Sonnet 65)

Shakespeare's treatment of this topic is indebted to a text that
has already been mentioned in the current chapter, namely Ovid's
Metamorphoses. In particular, he draws on the fifteenth and final
book, much of which consists of a speech by the Greek philosopher
Pythagoras (sixth century BCE) whose argument that 'All things do
change' offers an extended consideration of the theme that gives the
poem its title and subject. For example, the lines quoted above from
Sonnet 64 are modelled on the lines that Golding translates as 'I have
seen it sea which was substantial ground alate; | Again, where sea was
I have seen the same become dry land', while the opening lines of
Sonnet 60 recall another passage from Ovid:

> Like as the waves make towards the pebbled
>   shore,
> So do our minutes hasten to their end,
> Each changing place with that which goes
>   before;
> In sequent toil all forwards do contend.
>   (Sonnet 60)

> As every wave drives other forth, and that that
>   comes behind
> Both thrusteth and is thrust itself; even so the
>   times by kind
> Do fly and follow both at once and evermore
>   renew.[3]

Although it appears at the end of a poem whose central motif is the transformation of gods and humans into animals, plants, and minerals, Pythagoras's speech can legitimately be called scientific in its content and concerns, in so far as its speaker is described as teaching

> The first foundation of the world; the cause of
> everything;
> What nature was and what was God; whence
> snow and lightning spring;
> And whether Jove or else the winds in
> breaking clouds do thunder;
> What shakes the earth; what law the stars do
> keep their courses under;
> And whatsoever other thing is hid from
> common sense.[4]

It is difficult to be sure how to respond to Ovid's Pythagoras, whose belief that the souls of the dying transmigrate into other creatures is ridiculed in *Twelfth Night* (4.2.51–60) and whose consequent plea not to kill animals has 'effect but small' on his listeners; as Colin Burrow puts it, he is 'something of a wild-eyed vegetarian crank, whose philosophy can scarcely be taken straight'.[5] However, his treatment of matter as continually in flux noticeably underlies Shakespeare's imagery in the Sonnets, albeit with one important difference: while for Pythagoras, the fact that all things are continually changing is accompanied by a view of death as merely the prelude to the soul's transmigration to a new body, Shakespeare tends to emphasize transience and mortality more than renewal or rebirth. For example, Pythagoras's image of waves in motion, which he uses to make the point that 'the times … Do fly and follow both at once and ever more renew', is modified by Shakespeare into an image of waves making towards the shore, showing how 'our minutes hasten to their end'.[6] Similarly, while Pythagoras's recollection of seeing land being claimed by the sea is balanced by the recollection of sea becoming land, the lesson drawn by Shakespeare's speaker is one of transience: 'time will come and take my love away'. He responds much more noticeably to Pythagoras's characterization of time as 'the eater up of things'— 'Devouring time', as he calls it in Sonnet 19—than to the opposing insistence that nothing in the world perishes 'but, altering, takes new shape', and this allows him to allocate to himself the role of immortalizing the beloved in his 'black ink' (Sonnet 65).[7] The permanence that, in *Metamorphoses* 15, is paradoxically enabled through continual change is reconceived by Shakespeare as something to be achieved through poetry.

Ovid is far from being the only source Shakespeare draws on when exploring the transformation of matter in the Sonnets, however. A very different way of conceptualizing matter and what it can be made to do is referred to directly in Sonnet 114, where the speaker develops his complaint in the preceding poem that his eye transforms everything it sees—'The rud'st or gentlest sight, | The most sweet favour or deformed'st creature'—to images of his absent beloved. He goes on to speculate in 114 that 'your love taught it this alchemy, | To make of monsters and things indigest | Such cherubins as your sweet self resemble'. It might seem questionable to consider alchemy, a project concerned with the production of the Philosopher's Stone and the transmutation of base metals into gold, in the context of a book on science. However, over recent decades historians of science have attempted to reintegrate alchemy into their subject, both as an influence on the thinking of major figures such as Robert Boyle and Isaac Newton and for its importance in the evolution of scientific knowledge and technique. As William R. Newman explains, seventeenth-century chemists eager to distinguish their work from that of alchemists nevertheless used 'apparatus, practices, and skills developed by alchemists in trying to analyze plants and minerals', while alchemy 'guided the experimental ethos by stressing the need to subject matter to the artificial constraints of the laboratory'.[8] In Shakespeare's day, alchemy could be hard to distinguish from more conventional metallurgical projects. Queen Elizabeth's foremost advisor, William Cecil, financed unsuccessful attempts by the Venetian alchemist Giovan Battista Agnello and others to extract gold from ore brought back from the New World by Martin Frobisher in the late 1570s, as well as being a 'founding member of the Society for the New Art ... who promoted efforts to transmute base metals into copper'.[9]

The extent to which alchemy informs the Sonnets is difficult to determine with certainty. The word itself is used in Sonnets 33 and 114, in both cases as a metaphor (in 33, for the action of sunlight on water). Other parts of the sequence can be read as having alchemical overtones, including the opening of 119 ('What potions have I drunk of siren tears | Distilled from limbecks foul as hell within') and the procreation sonnets, in particular 5 and 6, with their recurrent imagery of distillation. (For a sixteenth-century image of alembics, see Figure 2.1.) However, some critics have argued that alchemical concepts are more pervasively present in the Sonnets than this: Margaret Healy, for example, sees Sonnet 19's lines addressed to Time, 'blunt thou the lion's paws … And burn the long-lived phoenix in her blood' as employing the terms alchemists used to refer to different stages in the process of making the Philosopher's Stone, and the marriage theme of Sonnets 1–17 as an allusion to the '"conjunctio" of opposites—of male and female, Sol and Luna, Sulphur and Mercury, fixed and volatile, form and matter—through their dissolution and blending by fire in the alchemist's alembic'. This need not imply that Shakespeare was interested in alchemy in any practical way: as Healy explains, alchemy in Shakespeare's time was understood both as scientific project and as an 'inner spiritual process'.[10] Instead, she argues that Shakespeare drew on a rich body of alchemic and magical

Figure 2.1. Image of alembics from Conrad Gesner, *The newe jewell of health*, tr. George Baker (London, 1576). Wellcome Collection.

writing to explore themes of love, inner transformation, and poetic creativity.

By its very nature, Healy's argument is a difficult one to verify: since writers on alchemy veiled their discussions under deliberately obscure language and imagery, we should not expect much direct treatment of alchemy in the Sonnets. However, the procreation sonnets, at least, do explicitly employ imagery that looks somewhat like alchemy when they exhort the fair youth to reproduce. The speaker of Sonnet 5 describes the depredations of winter on the plant life that flourished in summer, freezing its 'sap' and removing 'lusty leaves' from the trees:

> Then were not summer's distillation left
> A liquid prisoner pent in walls of glass,
> Beauty's effect with beauty were bereft,
> Nor it nor no remembrance what it was.
> But flowers distilled, though they with winter
>   meet,
> Leese but their show; their substance still lives
>   sweet.

Through distillation, the sweet aroma of summer flowers such as roses can be preserved even when their beautiful appearance has been lost. As Dympna Callaghan points out, the image is a domestic one: instructions for making rosewater can be found in early modern household manuals.[11] However, as Sonnet 6 continues the theme, Shakespeare's speaker applies the example to the fair youth:

> Then let not winter's ragged hand deface
> In thee thy summer ere thou be distilled.
> Make sweet some vial, treasure thou some
>   place
> With beauty's treasure ere it be self-killed.

Here, the youth is urged to take on the roles both of rose and of distiller, preserving his beauty against time's destructiveness through marriage and reproduction (the female partner in this enterprise being described with misogynistic reductiveness as a mere vessel or 'vial'). Again, the metaphor of distillation comes from perfumery or domestic husbandry; but its application to human bodies also has affinities with the ideas of the revolutionary Swiss physician Theophrastus von Hohenheim, also known as Paracelsus (1493–1541). Strongly influenced by alchemy, Paracelsus argued that all things were composed of mercury, sulphur, and salt, and he recommended the use of chemicals extracted from minerals and other substances in medical treatment. Paracelsus's view of 'the organs of the body in terms of alchemical apparatus, as so many stills, filters, casks, and the like' may inform the way Shakespeare imagines the making of semen as a process of distillation and likens the womb to a glass receptacle.[12] As Deborah Harkness explains, 'Paracelsian therapies were wildly popular among consumers' in Elizabethan London, as well as being a central point of contention in very public boundary disputes between physicians, surgeons, and unlicensed medical practitioners, so it is not fanciful to imagine that Shakespeare could have had this body of thought in mind.[13]

While Paracelsian medicine may provide some context for the distillation imagery in Sonnets 5 and 6, another aspect of their treatment of matter remains potentially confusing, and that is the use of the word 'substance' at the end of Sonnet 5 when the speaker says that distilled flowers 'Lose but their show; their substance still lives sweet'. As has already been discussed, when 'substance' is used in Sonnet 44 it refers to the speaker's material body, his 'flesh'. Here, however, it seems to refer to something much less tangible: the smell of the distilled rose, as well as any other properties that the distillation may have. In setting this 'substance' up in opposition to the flower's 'show', the speaker is saying that the growing flower—its colour, its shape, the way it feels to the touch, and so on—is only a shadow or image of its real substance, which is what gets preserved through distillation. The 'substance' of the rose persists, despite the fact that it no longer looks or feels like a rose, while the aspects of it that might seem most materially substantial are not where its true substance lies.

This is a paradox that Shakespeare plays with elsewhere in the Sonnets. In 37, for example, the speaker says that despite being bereft of good qualities, he takes all his comfort from the 'worth and truth' of the beloved; 'So then I am not lame, poor, nor despised, | Whilst that this shadow doth such substance give | That I in thy abundance am sufficed'. In asserting that a shadow can be substantial, Shakespeare is exploiting the multiplicity of meanings of the word 'substance' in English, including *OED* 6.b, 'That which gives a thing its character', 7.a, 'That of which a physical thing consists', and 11.a, 'Something that is solid or real (as opposed to and often contrasted with *appearance, semblance, shadow,* etc.)'. However, whether consciously or not, he may also have been influenced by contemporary debates over sense 4, 'The essential element underlying phenomena, which is subject to modifications', understood in sense 4.c, 'With reference to the doctrine of the Real Presence in the Eucharist'. Since 1215, the Catholic Church had held as canonical the doctrine of transubstantiation: the idea that in the sacrament of the Eucharist, the substance of the bread and wine that are offered is transformed by divine agency into the body and blood of Christ, while continuing to look, taste, and smell like bread and wine. Following the Reformation, however, Protestant churches rejected this doctrine both as lacking an unambiguous scriptural basis and as ascribing an unacceptable degree of spiritual agency to the clergy. While Protestant theologians differed considerably in their interpretations of what happens during the Eucharist, ranging from Martin Luther's notion of consubstantiation (in which the substance of the bread and wine remains present alongside that of Christ's body and blood) to Ulrich Zwingli's understanding of the Eucharist as a commemorative meal in which the body and blood of Christ were not present as substances, they invariably denied the possibility that the substance of the sacrament could be transformed.

Once again, this chapter might seem to be diverging from science into rival territory, namely religion: whatever people believe happens to bread and wine during the Eucharist, if their external properties stay the same then the topic can hardly be subjected to scientific analysis. However, the topic is relevant to this book, both because it led to debates over the properties of matter in which participants were prepared to die and kill for their beliefs, and because the doctrine of transubstantiation, rather than being seen purely as a matter of faith and beyond the realm of rational argument, had itself been formulated in terms derived from Aristotelian metaphysics. Medieval theologians, notably Thomas Aquinas, drew on the Aristotelian distinction between the substance of a body (what makes it what it is) and its accidents or incidentals to argue that although the Eucharistic offering is no longer bread and wine, the accidental properties of bread and wine remain in the space occupied by the substance of Christ's body and blood.[14] By separating substances from accidents in this special case, and by treating the dimensions of the bread and wine as accidents, Aquinas also explained how it was possible for Christ's body to be substantially present both in heaven and wherever Eucharist is being celebrated: 'Christ's extended body is in the sacrament as if it was just substance and not in the way a substance already envisaged as affected by quantity is underneath its quantity or in a place.'[15]

The logical complexities of the Catholic position were widely attacked by Reformation theologians and polemicists, not least for seeming to defy what today we would call the laws of physics. As John Foxe put it in his *Papa confutatus*, translated into English by James Bell (1580):

in respect of nature itself, what can be uttered more grossly absurd & more unreasonable, than to say that one and the self same Christ may sit in heaven, according to his natural and bodily presence, and at the self same time too be handled with hands on earth? that the substance of one self body is limited at one self time in one certain place, & to be contained in infinite places…?

This debate over the nature of the sacraments frequently employed terms like 'shadow', 'show', and 'substance' to convey its central ideas, such as where Foxe summarizes the Catholic position as being that 'under those shadows and accidents of bread and wine the very substance of flesh and blood is received'.[16] It made itself felt in plays like the mid-century *Jack Juggler* (which ridicules the idea that 'one man at a time may be in two places') and the work of religious poets from both sides of the confessional divide; given the pervasive nature of this debate, it is plausible to see it as informing poems like Sonnet 37, where the 'shadow' of the beloved 'doth such substance give | That I in thy abundance am sufficed | And by a part of all thy glory live'.[17] In claiming that a shadow can convey a substance that fills the recipient with glorious abundance, Shakespeare's speaker puts the terminology of Eucharistic debate to secular purposes. A sense of proportion has to be maintained here: Shakespeare's Sonnets are not primarily, or even especially, concerned with this theological question, pressing though it was throughout his lifetime. However, it can be seen as informing moments in the sequence when matter is imagined as doing impossible things, as when in 44 and 45 the speaker thinks of himself as being in two places simultaneously and fantasizes about a change in the 'substance of my flesh', or in the question that begins Sonnet 53, 'What is your substance, whereof are you made, | That millions of strange shadows on you tend?'. Although the theme is pursued in a Platonic direction, where all instances of natural beauty are mere shadows of the ideal that is the speaker's beloved, the idea that a single substance can generate millions of shadows parallels the question of precisely in what sense Christ is simultaneously present in heaven and in multiple locations where Eucharist is being celebrated.

It is not only in Shakespeare's Sonnets that an interest in the properties of matter is accompanied by imagery derived from Eucharistic controversy. A notable example appears in *Hamlet* when Claudius asks the Prince about the location of Polonius (knowing full well that Hamlet has killed him) and is told that he is 'at supper' (4.3.18):

> HAMLET   Not where he eats, but where a is eaten. A
> certain convocation of politic worms are e'en at him.
> Your worm is your only emperor for diet. We fat all
> creatures else to fat us, and we fat ourselves for
> maggots. Your fat king and your lean beggar is but
> variable service—two dishes, but to one table. That's
> the end.
> KING   Alas, alas!
> HAMLET   A man may fish with the worm that hath eat
> of a king, and eat of the fish that hath fed of that
> worm.
> KING   What dost thou mean by this?
> HAMLET   Nothing but to show you how a king may go
> a progress through the guts of a beggar. (20–31)

Hamlet's joking allusion to the 1521 Diet (or convocation) of Worms to which Martin Luther was summoned to justify his views establishes a context of theological controversy that, as Stephen Greenblatt has argued, shapes the meaning of his subsequent image of a beggar digesting a king. Opponents of the doctrine of transubstantiation made much of the physical indignities to which it seemed to subject the body of Christ, the King of Kings, in the form of the Host: 'If God was actually bread, they tirelessly jested, it meant that God could be eaten by worms, flies, or mice, that the divine body could decay and rot, or that, passing through the intestines, it could be transformed into excrement.'[18] Here, Hamlet's rather heartless insistence on the way death has transformed Polonius from live human to edible matter, and on worms' lack of respect for social distinctions, gains transgressive force from his appropriation of imagery derived from religious polemic. However, while Greenblatt is certainly correct to see the traumas and dissentions attendant on religious change as an important underlying concern of *Hamlet*, the play also displays a wider interest in matter and its properties. As Carla Mazzio puts it, the play has been 'mined, indeed gutted, in recent years for its preoccupation with the essence, or quintessence, of dust: dirt, land, and soil; flesh, bone and brain; worms, viscera, and human remains—in short, matter'.[19] As her list of nouns implies, what this amounts to is arguably a preoccupation with death: with the dust to which all creatures must return, with fleshy matter and its vulnerability, with decay and its agents. In so far as Shakespearean tragedy is concerned with death, it is inextricably concerned with the transformation of humans into inanimate matter, whether that be Cordelia, 'dead as earth' (*History of King Lear*, 5.3.257), Cleopatra 'marble-constant' (5.2.236), or the Ghost in *Hamlet*, which gruesomely recounts the hardening of Old Hamlet's body as Claudius's poison took its effect:

> swift as quicksilver it courses through
> The natural gates and alleys of the body,
> And with a sudden vigour it doth posset
> And curd, like eager droppings into milk,
> The thin and wholesome blood. So did it
> mine;
> And a most instant tetter barked about,
> Most lazar-like, with vile and loathsome crust,
> All my smooth body. (1.5.66–73)

The section that follows focuses on matter as it is imagined in three Shakespearean tragedies, *Hamlet*, *King Lear*, and *Antony and Cleopatra*. More specifically, it considers these plays in relation to a theory of matter that would become increasingly influential over the course of the seventeenth century, namely atomism.

Tragedy and Atomism

As with other branches of the sciences, the Aristotelian conception of matter that had informed medieval thinking continued to be dominant well into the seventeenth century.[20] Textbooks like the *Physica peripatetica* of Johannes Magirus (1597) and the *Corps de philosophie* of Scipion du Pleix (1645) relied on modified versions of Aristotelian doctrines, asserting that everything on earth was composed of the four elements with their associated qualities of heat, cold, dryness, and moisture. Like Aristotle, they treated physical entities as deriving from 'unformed matter' given tangible existence by an actuating principle of form. However, as Robert Kargon explains, 'This system of substantial forms and qualities was only with difficulty applied to physical problems by scientists in the late sixteenth and early seventeenth centuries.' Aristotelian physics could not satisfactorily explain the behaviour of magnets, for example, or the creation of a vacuum—something Aristotle believed was impossible—in Evangelista Torricelli's experiments with mercury in the 1640s.[21] Faced with phenomena like these, early modern scientists turned to other ways of thinking about matter and its behaviour, and one of these was offered by the atomist theory developed by the pre-Socratic philosophers Leucippus and Democritus (fifth century BCE) and, later, Epicurus (third century BCE). According to this tradition, as Daniel Garber summarizes it,

the world was made up of two principles: atoms and the void. Atoms were taken to be the smallest parts of matter, possessed of size, shape, weight, and nothing else. Although finite in size, and thus having physical parts, atoms were taken to be indivisible. In this way, they constituted the smallest level of analysis for any body. Furthermore, all the manifest properties of bodies were to be explained in terms of the size, shape, and motion of these atoms.[22]

Atomism was thus an important influence on some practitioners of the 'mechanical philosophy' of the seventeenth century (in particular, the Frenchman Pierre Gassendi, 1592–1655), who sought to explain the behaviour of material bodies in terms of the 'size, shape, and motion' of their parts.

The original writings of the ancient atomist philosophers have not survived, but their ideas were conveyed to early modern thinkers via sources including the works of Aristotle (who disagreed with them), the *Lives and Opinions of Eminent Philosophers* by Diogenes Laertius (third century CE), and *De rerum natura* (*On the Nature of Things* or *On the Nature of the Universe*), a six-book poem by Titus Lucretius Carus (first century BCE) that offers a detailed exposition of Epicurean beliefs. Praised in its own time by the Roman orator Cicero, this text was considered lost during the Middle Ages, although scholars such as Petrarch were aware that it had existed and had access to fragments preserved in other texts.[23] Following the discovery of a manuscript in a monastery in southern Germany by the humanist Poggio Bracciolini in 1417, the Latin text was widely copied, with the first of many printed editions appearing in 1473.[24]

De rerum natura was evidently a poem that interested major literary figures of Shakespeare's day: Ben Jonson had a copy, and that of the French essayist Michel de Montaigne was identified in the 1980s.[25] However, it also influenced some of the foremost scientists of the sixteenth and seventeenth centuries, to a degree that may seem surprising to the modern reader. Gassendi 'quoted roughly 5,300 of the 7,400 lines from *De rerum natura* over the course of his career', while the notebooks of Isaac Newton from 1664 to 1665 show him 'grappling with a proof for the existence of atoms and the void and thinking through the poet's arguments'.[26] Earlier, during Shakespeare's lifetime, Lucretius's exposition of Epicurus was an important influence on Francis Bacon and on the scientist and mathematician Thomas Hariot, among others.[27] As Kargon suggests, one reason for this may have been the ongoing dominance of the classical tradition: even when rejecting Aristotle, early modern thinkers sought legitimation from another ancient authority.[28] However, the sheer imaginative energy of the examples Lucretius uses to prove his atomist hypothesis, and their recourse to the details of everyday lived experience, also seem to have been compelling. A number of experiments by early modern scientists had their origins in Lucretius, such as Daniel Sennert's demonstration of the small size of invisible particles through making 'a stream of alcohol vapor pass through a sheet of writing paper that had been folded four times'.[29]

It's impossible to do justice to the richness, variety, and intellectual challenge of Lucretius's poem in a brief summary. However, some recurrent themes that are particularly relevant to the current chapter are as follows:

Everything in the universe is of 'fixed specific seeds' (1.169), which are 'everlasting' (1.221), although individual organisms are mortal. Although these particles are invisible, their existence can be inferred from the way smells reach our nostrils (1.298), clothes dry in the sun (1.305-10), and statues' hands get imperceptibly smaller by rubbing (1.316-21). However, there is also 'void in things' (1.330) which, for example, explains the varying densities of objects.

Those who think the world was created by a god out of nothing are wrong (2.167–183). 'Nothing can ever be reduced to nothing | Nor anything again grow out of nothing' (1.857–858). The existence of the world is due to chance:

> many primal atoms in many ways
> Throughout the universe from infinity
> Have changed positions, clashing among
>   themselves,
> Tried every motion, every combination,
> And so at length they fall into that pattern
> On which this world of ours has been created.
>   (1.1024–1028)

The gods are indifferent to human fortunes: 'They are not won by gifts nor touched by anger' (1.49). There is no such thing as Providence (2.167–183), and scary phenomena like thunder and lightning have natural, not divine causes (6.80–422). An understanding of this helps banish fear of the unexplained (1.151–154) and needs to be 'firmly founded in our senses' (1.423). Furthermore, the spirit is material and dies with the body (3.576–579), so there is no need to fear life after death: 'when the end shall come … nothing can harm us | Or make us feel, since nothing of us remains' (3.838–841). The 'tight knots of religion' are something from which our minds need to be 'set free' (1.931–932).

Unsurprisingly, the religious orthodoxy of Shakespeare's day found Lucretius, Epicurus, and atomism in general intolerable. While not, strictly speaking, atheistic, Lucretius's poem posited gods who were so indifferent to human concerns that it might as well have been.

It argued that the world had been created by chance rather than by design, denied the existence of Providence, and treated religion as a combination of fantasy and fraud; conversely, it asserted that true happiness consisted in understanding the workings of nature, in accepting that the soul was mortal, and in consequently losing the fear of punishment after death. As a result, 'epicure' in English tended to have pejorative connotations, indicating godlessness and/or a devotion to physical self-indulgence. Thomas Hariot's interest in atomist theory was treated as evidence of atheism in 1603 during the trial for treason of his patron Walter Ralegh, who was exhorted by Lord Chief Justice Popham 'Let not Hariot, nor any such Doctor, persuade you there is no eternity in Heaven, lest you find an eternity of hell-torments'.[30] The reference to 'any such Doctor' is perhaps meant to align Hariot with the notorious Doctor Faustus, who in Marlowe's play insists that 'hell's a fable'.[31]

Shakespeare was certainly aware of these negative connotations of Epicureanism. The first printed text to refer to Shakespeare by name, *Palladis Tamia* by Francis Meres (1598), notes in the same chapter that 'our tragical poet *Marlowe* for his Epicurism and Atheism had a tragical death', while Shakespeare himself treats 'epicurean' as a byword for gluttony in *Antony and Cleopatra* (2.1.24, 2.7.51), *King Lear* (*History*, 1.4.238) and *Macbeth* (5.3.8).[32] However, he was also evidently able to draw upon a more sophisticated understanding of Epicurean philosophy, such as when he has Cassius in *Julius Caesar* ponder whether, contrary to its tenets, inauspicious events may be portentous: 'You know that I held Epicurus strong, | And his opinion. Now I change my mind, | And partly credit things that do presage' (5.1.76–78). This raises the possibility that Shakespeare may have known Lucretius's poem directly, perhaps through Jonson; certainly, at some point in his career he became familiar with a text in which *De rerum natura* is frequently quoted, namely Montaigne's *Essays* (printed in English 1603 but possibly available to Shakespeare via their translator John Florio before then). Montaigne's fascination with Lucretius is evident from the *Essays*: in a text saturated with quotations from classical authors, he quotes *De rerum* nearly 150 times and repeatedly engages with Epicurean philosophy—most extensively in the 'Apology for Raymond Sebond', which includes a detailed (and critical) discussion of the idea that the world is the product of chance collisions of atoms.[33] Shakespeare could easily have learned a great deal about Epicureanism without actually reading Lucretius.

This makes the search for Lucretian influence on Shakespeare a rather uncertain business, exacerbated by the fact that no English translation was available in his lifetime (in contrast to Ovid, where the indebtedness of Prospero's speech renouncing magic in *The Tempest* to Medea's incantation in *Metamorphoses* book 7 can readily be demonstrated by examining Golding's version). The term 'atomy', which appears in *As You Like It* 3.2.227 and 3.5.13, is derived from Epicureanism, but is used here simply to mean 'a mote in a sunbeam' (*OED* 'atom', *n.*, 7) rather than calling up a specifically Epicurean frame of reference. It is used in a similar sense by Mercutio in *Romeo and Juliet* when describing Queen Mab:

> Drawn with a team of little atomi
> Over men's noses as they lie asleep.
> Her wagon spokes made of long spinners' legs,
> The cover, of the wings of grasshoppers ...
>   (1.4.58–61)

Suparna Roychoudhury links this to the way Lucretius helps his reader to imagine the existence of atoms by making reference to 'animals so small | That a third part of them is quite invisible' (4.116–117); furthermore, Lucretius's attribution of dreams to 'images of things' that 'wander about | In all directions' (4.724–776) before penetrating the sleeping mind is analogous to Mercutio's fantastic notion that dreams have their origins in something material, albeit minuscule, outside the dreamer.[34] While the lack of a close verbal correspondence makes this argument hard to prove, the speech does testify to an interest in the microscopic in Shakespeare's plays of the 1590s: both *Love's Labour's Lost* and *A Midsummer Night's Dream* include characters named Moth (or 'Mote'). However, one argument for Shakespeare's encountering *De rerum natura* directly around the turn of the century may be the fact that after *As You Like It*, the term 'atomy' never reappears in his work. As Jonathan Pollock points out, 'Lucretius does not call Epicurus's *atomoi* "atoms", but "grains" and "seeds": *semina rerum*', and Shakespeare uses similar terminology in *Measure for Measure* and

King Lear.[35] In the former, the Duke, disguised as a friar, tries to reconcile Claudio to the prospect of death in incongruously secular terms, treating death as oblivion rather than invoking the prospect of an afterlife:

> Thy best of rest is sleep,
> And that thou oft provok'st, yet grossly fear'st
> Thy death, which is no more. Thou art not
>   thyself,
> For thou exist'st on many a thousand grains
> That issue out of dust. (3.1.17–21)

As Robert N. Watson puts it, the Duke's argument to Claudio in the latter part of this quotation is that 'from the standpoint of atomistic philosophy, he really has no self to lose': Claudio consumes atoms in the form of food, he converts those atoms into the stuff of his own body, and, by implication, he will be dispersed back into atoms on his death.[36] Pollock's second example comes from the King's speech to the storm in *King Lear*:

> and thou all-shaking thunder,
> Smite flat the thick rotundity of the world,
> Crack nature's mould, all germens spill at once
> That make ingrateful man. (*History*, 3.2.6–9)

Although Lear's plea to the thunder is inherently un-Lucretian in its implicit assumption of supernatural entities who listen to human requests, his idea of man as made of 'germens' or seeds is potentially atomistic, although another possible source is Ovid's suggestion that the first men were created from 'heavenly seed' (1.92). However, the case for a Lucretian frame of reference is strengthened by Lear's response to Cordelia's defiant 'Nothing' in the opening scene. 'Nothing can come of nothing' (1.1.82), though a proverbial phrase, was also a key atomist tenet, as exemplified by the way John Aubrey used it to describe Thomas Hariot's death from a tumour that began as a speck on the nose: 'he would say *ex nihilo nihil fit*. But ... a *nihilum* killed him at last'.[37] Mary Thomas Crane uses these possible references to atomism as her justification for reading *King Lear* in relation to '[e]arly modern debates about the nature of matter'; specifically, she links the atomist argument that at a microscopic level matter is very different from how it is apprehended by the human senses to the play's 'treatment of epistemological and ethical questions' such as whether a monarchy can be divided, whether a blind man can know when he is ascending a slope or falling through the air, and whether filial love can be quantified.[38]

If Shakespeare was indeed influenced by Lucretius when writing *King Lear*, it has to be admitted that his treatment of Epicureanism falls some way short of unqualified acceptance. Lear's use of the 'nothing can come of nothing' tag appears during a moment of catastrophic misjudgement, while the character who voices Epicurean attitudes most fully is the amoral Edmund, who in his first soliloquy celebrates nature and who shortly afterwards disclaims the influence of heavenly bodies on his personality: compare Lucretius's repeated elevation of nature above the gods (e.g., 2.1092), his rejection of supernatural explanations, and his emphasis on free will (2.256). Edmund's betrayal of brother and father seems to bear out the Protestant theologian Philipp Melanchthon's complaint that 'Since violence often rules, and there are many unforeseen changes, the Epicureans believe that human affairs are moved by chance, and that each can obtain as much as he is worth in power.'[39] In *Hamlet*, however, the thinking through of the implications of atomist ideas is less beset by moral anxiety. The Prince's enquiry of the Gravedigger as to what happens to the body after death seems, rather, to be expressed in a spirit of morbid curiosity: 'How long will a man lie i'th' earth ere he rot?' (5.1.159). The question prompts a comparison of the durability of the corpses of tanners and the pox-ridden, and of course the Gravedigger's production of the skull of 'poor Yorick', a graphic testimony to posthumous decay: 'Here hung those lips that I have kissed I know not how oft' (183–184). However, while Yorick's lips, along with the rest of his flesh, have disappeared, Hamlet's train of thought insists on the durability of the matter from which they were composed:

> HAMLET   Dost thou think Alexander looked o' this
>   fashion i'th' earth?
> HORATIO   E'en so.
> HAMLET   And smelt so? Pah!
> HORATIO   E'en so, my lord.

> HAMLET To what base uses we may return, Horatio!
> Why may not imagination trace the noble dust of
> Alexander till a find it stopping a bung-hole?
> HORATIO 'Twere to consider too curiously to consider
> so.
> HAMLET No, faith, not a jot; but to follow him
> thither with modesty enough, and likelihood to lead
> it, as thus: Alexander died, Alexander was buried,
> Alexander returneth into dust, the dust is earth, of
> earth we make loam, and why of that loam whereto
> he was converted might they not stop a beer-barrel?
> Imperial Caesar, dead and turned to clay,
> Might stop a hole to keep the wind away. (193–209)

The perception that drives Hamlet's imagination here is that while everything dies, matter is eternal. This is the flip-side to Lear's remark that nothing can come of nothing: in Lucretius, remember, that doctrine is preceded by the line, 'Nothing can ever be reduced to nothing.' The indestructible atoms that composed Alexander and Caesar are still out there somewhere. Sean Ferrier and Lisa Walters characterize Hamlet's emphasis on the 'materiality of death' as Lucretian, and certainly, Lucretius seems to have prompted this kind of reflection in his early modern readers: Montaigne wrote (in Latin) on a flyleaf of his copy of *De rerum natura*, 'Since the movements of the atoms are so varied, it is not unbelievable that the atoms once came together, or will come together again in the future, so that another Montaigne be born.'[40]

In the playhouse, Hamlet's discussion of the afterlife of matter may gain an additional charge as a result of the performance practices of the early modern theatre. As William N. West points out, in Shakespearean drama questions of matter, identity, space, and time 'emerge as quintessentially theatrical problems, embodied onstage in the recurring tropes of disguise, or of two bodies who claim the same name, or of indistinguishable twins, even of the ordinary staging practices like the doubling of parts or the transvestite playing of women by boys'. The protean actor can be understood as unformed matter who can transform himself into different shapes through his own technical skill.[41] The theory of matter that West draws on here is primarily Aristotelian, not atomist, and he does not develop in detail the point about doubling made at the start of his essay. However, his insight is relevant to the graveyard scene of *Hamlet* because of the potential for the part of the Gravedigger to be doubled with that of Polonius—an obvious way of reducing the number of actors, given that Polonius dies in Act 3, and one whose known history stretches back to 'at least 1730'.[42] This gives an unsettling comic edge to the Gravedigger's reply to Hamlet's question, 'Whose grave's this, sirrah?', 'Mine, sir' (5.1.115–116), as if it is the dead Polonius who is speaking. It also means that Hamlet's speculations about the persistence of matter after death acquire an uncanny theatrical analogue in the actor playing the Gravedigger, a reminder that after death bodies do not simply disappear—something that has already been made graphically clear through Hamlet's having to 'lug the guts' of the dead Polonius into the next room (3.4.186), a recognition of the perennial theatrical problem of getting corpses offstage. The afterlife of Polonius as Gravedigger parallels the afterlife of Alexander as stopper; and if, as Polonius's reminiscences of playing Caesar at university imply, the same actor played the title role in Shakespeare's own *Julius Caesar*, then Hamlet's evocation of 'imperial Caesar' at this point generates a further intertheatrical effect.[43] The graveyard scene in *Hamlet* engages with matter theory in a way that is shaped by the affordances of the theatrical medium itself.

Antony and Cleopatra extends *Hamlet*'s interest both in the inevitable slide of sentient humans into inert matter, and in the capacity of matter to generate and sustain new life. This preoccupation is facilitated by the play's setting: the flood-plain of the Nile was widely believed to bring forth tiny animals by spontaneous generation when warmed by the sun, a notion that Cleopatra refers to when she swears 'By the fire | That quickens Nilus' slime' (1.3.68–69). Shakespeare could have encountered this in the *Metamorphoses*, when Ovid uses it as a point of comparison for the resurgence of new life after Deucalion's flood: when 'the sun did heat the mud the which he left behind, | The husbandmen that tilled the ground among the clods did find | Of sundry creatures sundry shapes' (1.505–507). But it also exemplifies a very Lucretian preoccupation with the cyclic transformation of matter, by which faeces generates plant and animal life: 'Our dungy earth alike | Feeds beast as man' (1.1.37–38). Cleopatra registers an awareness of the persistence of matter through the food chain when determining on suicide:

> And it is great
> To do that things that ends all other deeds,
> Which shackles accidents and bolts up
> change,
> Which sleeps and never palates more the
> dung,
> The beggar's nurse, and Caesar's. (5.2.4–8)

Although what she desires is, in Lucretian terms, impossible, in that there is no end to the journey of matter, her understanding that food, the body, and earth are all essentially dung does closely parallel Lucretius's explanation of how even living things are 'born from insentient atoms':

> Why, you can see that living worms emerge
> From filthy dung when the wet earth is soaked
> And rotted by unseasonable rains.
> All other things are seen likewise to change.
> Rivers and leaves and joyful pastures change
> Into cattle, and cattle change into our bodies,
> And often too our bodies build the strength
> Of wild beasts and winged masters of the air.
>   (2.870–878)

That final recognition that human bodies are, themselves, food for other creatures seems to inform Shakespeare's depiction of Cleopatra's own end. The asp that the Clown brings her in 5.2 is repeatedly referred to as a 'worm', a word that never occurs in Shakespeare's source, Thomas North's translation of Plutarch's account of her death, and one effect of this is to elide the difference between the creature that dispatches her and the organisms that will later consume her. Indeed, after applying the asp she describes it not just as poisoning her, but as feeding: 'Dost thou not see my baby at my breast, | That sucks the nurse asleep?' (304–305). Materially, something of her has passed into the worm, and this makes its disappearance before the

arrival of Caesar's party (who can discern only 'an aspic's trail') a synec-
doche for her own outwitting of her enemy. Like the asp—and in the
asp—she has already left the building.

The Tempest

A final Shakespearean episode that has been read as a possible sign of
Lucretian influence appears in *The Tempest* in the form of Miranda's
response to the shipwreck with which the play begins. At the start of
the second book of *De rerum natura*, Lucretius uses the idea of a storm
at sea as a way of conveying the calmness of spirit that should attend
those who have rejected the struggle for wealth and power:

> A joy it is, when the strong winds of storm
> Stir up the waters of a mighty sea,
> To watch from shore the troubles of another.
> No pleasure this in any man's distress,
> But joy to see the ills from which you are
> spared ... (2.1–4)

This passage was frequently cited: to give two instances Shakespeare
may well have been familiar with, Montaigne quotes the passage at
the start of his essay 'Of profit and honesty', and a translation of it
appears in the highly influential collection of lyrics, *Tottel's Miscellany*
(1557). In *The Tempest*, however, while Miranda and Prospero watch
the storm from a place of safety, Miranda's empathetic response is
hardly the feeling of relief imagined by Lucretius: she says she has
'sufferèd | With those that I saw suffer' (1.2.5–6), expressing a feeling
of painful sympathy rather than calm detachment.[44] The play's treatment of matter, too, avoids the Lucretian implications detectable in
Hamlet, *King Lear*, and *Antony and Cleopatra*. It seems to be carefully
structured along elemental lines, with Prospero setting up a verbal
contrast between the 'Delicate Ariel' (1.2.444), rarefied and often
invisible, and the 'earth' (316) Caliban, while the elements themselves,
rather than being subject purely to natural forces, are open to magical
control. In his valedictory speech to the spirits who have helped him,
Prospero says he has 'called forth the mutinous winds, | And 'twixt
the green sea and the azured vault | Set roaring war'; 'to the dread

rattling thunder … given fire'; and the 'strong-based promontory | … made shake' (5.1.42–47).

Prospero's impressive claims here identify him more as a magician than as a scientist, and a great deal of critical energy has been expended on working out what kind of magician: a 'Renaissance philosopher-magus' along the lines of the German philosopher Heinrich Cornelius Agrippa; a witch, as is implied by Shakespeare's use of an incantation by Ovid's Medea when writing the speech quoted above; a product of literary tradition; or indeed a charlatan.[45] This does not bar him from consideration in a scientific context: as with alchemy, historians of science regularly take account of magic, not simply as a misguided belief system that later ages would discard, but as a collection of ways of making sense of unexplained phenomena that paralleled scientific enquiry, were hard to disentangle from it, and were frequently practised by the same people.[46] John Dee, for example, was both an accomplished mathematician whose 'expertise encompassed astronomy, astrology, navigation, and mechanics', and an occult philosopher who believed he was communicating with angels via his intermediary Edward Kelley to whom they 'dictated entire books, written in an angelic language that only Kelley could understand'.[47] The distinction between, say, atomism, which posits the existence of invisible particles to explain natural phenomena, and occult philosophy, which posits a hidden sympathy between celestial forces and objects on earth, is also perhaps less clear-cut than hindsight might like it to be. However, another way of reading Prospero's magic might be as an oblique exploration, on Shakespeare's part, of situations, problems, and power relations attendant on early modern science. One of his dramatic forerunners, the first title character in Robert Greene's play *Friar Bacon and Friar Bungay* (c. 1590), is, like Prospero, a magician who with the aid of supernatural beings is able to create spectacular entertainments, transport bodies through space, and see what is being done in distant places. However, the historical Roger Bacon was—as Greene would have well known—not a necromancer but a natural philosopher, important for his work on optics, who far from conjuring spirits believed that 'many effects that might vulgarly be accounted magical and dependent upon demons were in fact perfectly legitimate products of art and nature'.[48] As Robert Recorde wrote of Bacon in his geometry textbook *The Pathway to Knowledge* (1551):

Great talk there is of a glass that he made in Oxford, in which men might see things that were done in other places, and that was judged to be done by power of evil spirits. But I know the reason of it to be good and natural, and to be wrought by geometry (sithe [since] perspective is a part of it) and to stand as well with reason as to see your face in common glass.[49]

Greene's more fanciful depiction of Bacon as a raiser of devils exemplifies how the familiar dramatic figure of the magician could offer a way of representing diverse, and not necessarily magical, types of knowledge of and power over nature.

The possibility of reading Prospero not simply as a magician but as some kind of scientist has been a productive one for critics who have linked him, not just with magi like Agrippa and Dee, but with the more conventionally scientific figure of Francis (as opposed to Roger) Bacon. Bacon himself believed that 'the misapplied and abused Name of NATURAL MAGIC, which in the true sense, is but NATURAL WISDOM', might more properly be used to refer to the 'OPERATIVE' branch of natural philosophy, 'the PRODUCTION OF EFFECTS'.[50] Elizabeth Spiller and Mickaël Popelard have both drawn parallels between Bacon's insistence on the importance of putting scientific knowledge to useful ends and the way exile forces Prospero to move from 'secret studies' (1.2.77) to the operative use of his magical arts to achieve material goals (in his case, the controlling of Ariel and Caliban, the humiliation of his enemies, the regaining of his dukedom, and the installation of his future descendants on the Neapolitan throne).[51] Spiller also notes Bacon's influential belief in the validity of knowledge obtained under artificial conditions—a departure from scholastic traditions that privileged knowledge based on observing the ordinary course of nature. Although she does not use the word, she sees Prospero's island as a kind of laboratory, a 'space of experiment' in which Miranda can be brought up in isolation and Ferdinand introduced to her.[52]

If the island is an experimental space, though, that does not make it a politically neutral one: as Caliban forcefully insists, 'This island's mine, by Sycorax my mother, | Which thou tak'st from me' (1.2.332–333). The space in which Prospero can carry out a

Baconian 'transformation of nature', to use Popelard's term, is one that has been wrested from its existing inhabitant (or inhabitants, if one includes Ariel). The play's interest in the practical uses of secret knowledge, be that scientific or magical, cannot be separated from its interest in colonization: as Denise Albanese argues, 'it is as though the colonialist aspects of location itself enable book-bound magic to be turned into a species of science'.[53] Prospero's ability to put his learning into effect—something he conspicuously failed to do in Milan—seems to be facilitated by the way exile presents him with a near-empty space and inhabitants susceptible to his control. The way Prospero's access to magic creates a power differential between him and Caliban invites comparison with the different technologies available to Europeans and to the people they encountered in the New World; one example especially pertinent to the current chapter appears in *A Brief and True Report of the New Found Land of Virginia* (1588), in which Thomas Hariot describes the flora, fauna, and people he encountered in 1585–1586 while surveying the land and ascertaining its appropriateness for trade and settlement. Having earlier described how Algonquian people dress in deer skins and have 'no edge tools or weapons of iron or steel to offend us withal', Hariot goes on to describe their reaction to being shown European artefacts:

Most things they saw with us, as Mathematical instruments, sea compasses, the virtue of the loadstone in drawing iron, a perspective glass whereby was shewed many strange sights, burning glasses, wildfire works, guns, books, writing and reading, spring clocks that seem to go of themselves, and many other things that we had, were so strange unto them ... that they thought they were rather the works of gods than of men, or at the leastwise they had been given and taught us of the gods.[54]

In Hariot's account, the Algonquians' inability to explain the workings of the unfamiliar technology they are shown leads them to posit a supernatural explanation: in effect, the distinction of science from magic depends on whether one understands its underlying principles.[55] (Ironically, when presented with a phenomenon he cannot explain—the mass death of native people from an illness that appears to leave the English untouched—Hariot himself is in the position of suggesting divine causes: 'some said that it was the special work of

God for our sakes, as we ourselves have cause in some sort to think no less'.)[56]

However, while Prospero, in his possession of technical knowledge unavailable to Caliban, may have affinities with the English would-be colonists of Virginia, and while his magical control over the elements may resemble the ambitions of scientists like Francis Bacon, a reading of the play that identifies him as a quasi-scientist is complicated by the nature of what he actually achieves in the course of it. Although his valedictory speech has him calling up winds, generating thunderbolts, creating earthquakes, and raising the dead, his actions during the play are of a different order. They include using Ariel to create fiery effects that cause the Italians to abandon ship; having the invisible Ariel make music, sometimes with magical results (for example, sending some of Alonso's party to sleep); immobilizing Ferdinand; producing a banquet which Ariel, in the shape of a harpy, makes disappear; generating a masque-like entertainment for Ferdinand and Miranda; having Ariel lead Caliban, Stephano, and Trinculo into a pond, after which they are chased by spirits in the shape of dogs; charming Alonso's party into a magic circle, and having Ariel bring the Master and Boatswain; and, in anticipation, having Ariel produce good weather to speed the Italians' progress back to their fleet. Not only is much of this done by Ariel rather than by Prospero directly; much of it consists of illusion rather than direct manipulation of natural bodies. Prospero, moreover, has 'Fortune' to thank for bringing his enemies to the island, and his success is time-sensitive, depending on 'a most auspicious star' and actions carried out ''twixt six and now'. His magic is both surprisingly limited in some respects, and akin to the workings of theatre in its reliance on disguise, special effects, and music.

If Prospero's magic is indeed fundamentally theatrical or illusionistic, then that has implications for a reading of the play that would identify him as a scientist under the guise of a magus. It might suggest a scepticism as to the claims of science to be able to change or improve on nature; the world of early modern science was characterized by charlatans claiming non-existent achievements, such as the Dutch engineer Cornelius Drebbel who in the years leading up to *The Tempest* was notorious in London for a machine that he said worked in perpetual motion.[57] However, a more positive interpretation might be to focus on the power of theatre to imagine and depict marvels with which scientific reality had not yet caught up, precisely because of its own ambiguously material status. Prospero's speech at the dissolution of his entertainment for Ferdinand and Miranda is relevant here, albeit somewhat misleading:

> These our actors,
> As I foretold you, were all spirits, and
> Are melted into air, into thin air;
> And like the baseless fabric of this vision,
> The cloud-capped towers, the gorgeous
> palaces,
> The solemn temples, the great globe itself,
> Yea, all which it inherit, shall dissolve;
> And, like this insubstantial pageant faded,
> Leave not a rack behind. We are such stuff
> As dreams are made on, and our little life
> Is rounded with a sleep. (4.1.148–158)

Although the references to 'actors' and 'the great globe' make these lines unavoidably metatheatrical, they are not straightforwardly applicable to Richard Burbage and his colleagues, who, far from being 'spirits', were made of flesh and blood and who—as the example of Polonius indicates—were hardly able to melt 'into thin air'. What *The Tempest* does exemplify, however, is the capacity for the early modern theatre to play games with materiality: to use the material to represent the immaterial. Ariel, for example, may tell the Italians that 'The elements | Of whom your swords are tempered may as well | Wound the loud winds, or with bemocked-at stabs | Kill the still-closing waters' as injure him (3.3.61–64), but the actor playing Ariel is in reality human and vulnerable, while the stage direction calling for 'divers spirits in shape of dogs and hounds' (5.1.253 s.d.) to pursue Caliban, Stephano, and Trinculo implies a complex process in which material actors pretend to be immaterial spirits pretending to be material animals.

The paradoxical nature of Prospero's speech, which gestures simultaneously towards airy spirits and earthly actors, highlights the tensions and contradictions that made early modern theatre as an art form peculiarly well placed to explore questions about the nature of matter:

not simply what stuff dreams are made on, but what stuff stuff is made on. On the one hand its reliance on the stubborn materiality of the actor's body, visible at moments such as Hamlet's awkward disposal of Polonius, emphasized the persistence of matter even after bodily death. On the other, though, its use of practices and tropes such as doubling and disguise, as West argues, was apt for the consideration of matter's openness to change. Jenny C. Mann and Debapriya Sarkar perceptively note that the shape-shifting god Proteus both served as 'a figure of the actor par excellence' in Renaissance culture and was used by Francis Bacon 'throughout his philosophical works' as a rhetorical figure for representing the instability of matter and the need to subject it to the discipline of experimental scrutiny.[58] It is appropriate, then, that Shakespeare's most literally protean character—the tricksy Ariel, who takes on shapes ranging from sea-nymph to harpy—should appear in a play that combines a thematic interest in human control of the elements, and the possible implications of that control, with an abiding and self-conscious metatheatricality.

3

The Body

The final sections of the last chapter showed how the magic Prospero uses to shape the action of *The Tempest* has affinities with the experimental science theorized by Francis Bacon, with technologies of colonial domination, and with the illusionistic spectacles of the early modern theatre itself. On an island that can be read variously as a place of exile, a nascent colony, a scientific laboratory, or a metaphor for the Shakespearean playhouse, Prospero uses his ambiguous powers to bring about a turn of events that combines tragicomic revenge (the regaining of his dukedom) with romantic comedy (the union of Ferdinand and Miranda). This gives his role in the play a metadramatic aspect: he deliberately creates obstacles to the plot's resolution, not least his own fictitious opposition to the lovers' match, in order to resolve them, a practice that recalls the machinations of Oberon in *A Midsummer Night's Dream* and Duke Vincentio in *Measure for Measure*.

In the context of a discussion of Shakespeare and science, however, another of Prospero's forerunners in the role of quasi-dramatist especially repays scrutiny. In *All's Well that Ends Well*, Helen demonstrates medical expertise in her curing of the King's fistula, thereby generating the very problem that she spends the bulk of the play resolving: an unwilling husband who refuses to acknowledge her as his wife until she has satisfied a seemingly impossible set of conditions. In order to fulfil these, she employs not only her native ingenuity but also, it can be argued, the same medical knowledge that she put into practice at the outset, engineering a final scene whose highly theatrical sequence of illusions and revelations forces Bertram into the

Shakespeare and Science. Tom Rutter, Oxford University Press. © Tom Rutter (2024).
DOI: 10.1093/oso/9780192898548.003.0004

position of accepting her. This makes Helen a good starting-point for a discussion of how early modern understandings of the body and its workings informed Shakespeare's writings, not just in their content but in their formal properties.

The opening scenes of *All's Well* revolve around two problems, one medical, the other romantic. The King of France, ailing with a fistula, has 'abandoned his physicians' (1.1.12) in despair at their inefficacy and longs only for death, lamenting that the one man with the skill to cure him, Gerard de Narbonne, recently died. Gerard's daughter Helen, meanwhile, consumed with desire for the son of her father's patron the Count of Roussillon, undertakes to cure the King, hazarding her life if she fails and requesting her choice of husband if she succeeds. His response to her initial offer reflects his assumptions about where medical expertise is to be found:

> We thank you, maiden,
> But may not be so credulous of cure,
> When our most learnèd doctors leave us, and
> The congregated College have concluded
> That labouring art can never ransom nature
> From her inaidible estate. I say we must not
> So stain our judgement or corrupt our hope,
> To prostitute our past-cure malady
> To empirics ... (2.1.113–121)

As far as the King is concerned, medical knowledge resides in the 'congregated College' of physicians, not with 'empirics' like Helen. Shakespeare's terminology here derives not from his source, the thirty-eighth novel in William Painter's *The Palace of Pleasure*, but the circumstances of early modern London, where the Royal College of Physicians had been founded in 1518 in large part as 'a vocational body, charged with the repression of unqualified practitioners'.[1] While surgeons were only meant to deal with the external parts of the body, and apothecaries were supposed to be limited to dispensing physicians' prescriptions, members of the College laid claim to a superior expertise in medicine based on their university education. As well as taking action against surgeons and apothecaries alleged to have gone beyond their remit, they sought to limit the business of unlicensed practitioners, a somewhat hopeless exercise in a 'swarming and disease-ridden city'.[2] The term 'empiric' was frequently used to disparage those whose knowledge derived from experience alone rather than theoretical understanding: as the physician John Cotta put it in an attack upon 'ignorant and unconsiderate practisers of physic', 'The Empiric ... only informeth himself by such things as oft appear evident & manifest unto sense and experimental proof'.[3] The King's answer to Helen, therefore, derives from contemporary debates over the value of different kinds of medical knowledge, and boundary disputes between those who possessed them. These disputes were inherently gendered, in that women were excluded from the university education that validated physicians' superior status. As such, Helen can by definition possess (as the King puts it) 'no art', notwithstanding the fact that (as Mary Lindemann observes) 'the majority of medical practitioners in the early modern world were women', working in the sphere of the household or local community.[4]

In fact, the nature and extent of Helen's medical knowledge are left somewhat ambiguous: as Mary Floyd-Wilson points out, 'what the cure involved, why it worked, and how it was effected' are a matter that 'remains occulted'.[5] Helen herself seems to disparage her abilities, telling the King that her father gave her 'Many receipts' (2.1.104) on his death bed and so presenting herself as merely a passive recipient of his written cures. She minimizes her agency by suggesting that God works through her: 'He that of greatest works is finisher | Oft does them by the weakest minister' (2.1.135–136). But she later goes on to insist, 'My art is not past power' (2.1.158), a phrase that expresses a sense of pride and ownership, while her request that the King accept 'the help of heaven' (2.1.152) jars a little with the lines from her soliloquy in the opening scene, 'Our remedies oft in ourselves do lie | Which we ascribe to heaven' (1.1.212–213). It is difficult to resist the conclusion that her disavowal of autonomous medical knowledge is strategic, a defensive manoeuvre intended to disarm the King's objections. Indeed, the pervasiveness of the assumption that a woman could not heal the King unaided becomes apparent after her success, which Lafeu attributes to the 'Very hand of heaven' (2.3.32):

They say miracles are past, and we have our philosophical persons to make modern and familiar things supernatural and causeless. Hence is it that we make trifles of terrors, ensconcing ourselves into seeming knowledge when we should submit ourselves to an unknown fear. (2.3.1–6)

The curing of the King by the 'weak' and 'debile' (2.1.34–35) Helen is here treated as a prodigious event that demands to be read as an instance of divine intervention in nature, like the startling meteorological events discussed in Chapter 1. This, too, prompts opportunistic publications purporting to identify Providence at work, 'A showing of a heavenly effect in an earthly actor' (2.3.24–25).

Whatever the King himself believes about the source of Helen's ability, he is sufficiently grateful to agree to her request for permission to marry Bertram—prompting Bertram's departure for the Italian wars and repudiation of their marriage until she can 'get the ring upon my finger, which never shall come off, and show me a child begotten of thy body that I am father to' (3.2.57–59). Helen's ultimate satisfaction of these conditions relies on an act of sexual substitution whereby, after faking her own death, she takes the place of the Italian maiden Diana with whom Bertram is besotted, exchanges rings with him in the process, and presents herself to him, pregnant, in the final scene. However, the fairytale nature of the plot can obscure the extent to which Helen's success relies, once again, on medical knowledge: in effect, she has to find a way of making sure that she gets pregnant from a single night's encounter. This kind of problem was of significant concern to some early modern physicians. In his widely published *Examen de ingenios para las sciencias*, for example (translated into English in 1594 as *The Examination of Men's Wits*), the Spaniard Juan Huarte spends considerable time explaining what couples should do to ensure that they have healthy, preferably male children, and emphasizes the importance of diet, exercise, timing of intercourse in relation to the menstrual cycle, and sexual position. Furthermore, the woman should rest, 'when the act of generation is ended, with her head down and her heels up'—on her right side, if she wants a boy.[6] Although it would be excessive to assume that Helen follows Huarte's advice in all its particulars, the fact that the opening scenes of the play hinge on her medical knowledge makes it likely that the denouement, too, is facilitated by a comparable expertise rather than mere good luck.

The instance of Helen in *All's Well that Ends Well* points towards a number of themes that are relevant to the current chapter. First, as has already been noted, the way in which characters like Lafeu question whether her healing of the King derives from her own abilities or divine intervention reflects not just sexism, but also a context in which natural explanations for diseases and cures coexisted with supernatural ones. Second, the way in which the plot of *All's Well* turns on medical expertise exemplifies a linkage between dramatic structures and early modern understandings of the body that can be seen in other plays—perhaps most notably, in so-called humours comedies. Finally, there is the way the play treats the female body—specifically, the female reproductive body—as marking the limits of medical knowledge. Presenting Bertram with his ring and the spectacle of herself with child, Helen asks him, 'Will you be mine now you are doubly won?' (5.3.316). His response, 'If she, my liege, can make me know this clearly | I'll love her dearly, ever ever dearly' (5.3.317–318), represents her pregnancy as a mystery the making-known of which constitutes yet another riddle—albeit one that may not be finally solvable, as that rather loaded word 'if' implies. The notion of the female body as the object of fascinated scientific scrutiny, and yet as perhaps fundamentally resistant to that scrutiny, is evident both in comedies like *All's Well* and, with more violent implications, in tragedies like *Hamlet* and *Othello*.

In exploring the way in which Shakespeare's work is informed by early modern conceptions of the human body and its workings, the current chapter addresses a topic that has generated so much and so varied criticism over recent decades as to defy brief summary. One important strand of criticism, exemplified in the work of Gail Kern Paster and Michael Schoenfeldt, has focused on the importance of the humoral model of physiology in early modern thinking. This venerable set of beliefs explained health and disease in terms of the balance of blood, phlegm, and yellow and black bile, and will be discussed in detail below.[7] Humoral balance was believed both to influence and to be influenced by the affections or passions, reflecting a wider set of assumptions about 'emotions as physical, environmental, and external phenomena', as Paster, Katherine Rowe, and Mary Floyd-Wilson put it, and the history of the emotions has been another productive area of Shakespeare criticism since the turn of the century.[8] Early modern conceptions of the body have also been of interest both to feminist critics and to others preoccupied with the ways in which beliefs about sex difference underpinned the workings of Shakespearean drama. Two influential texts in this regard have been the chapter on *Twelfth Night* in Stephen Greenblatt's *Shakespearean Negotiations* and Thomas Laqueur's *Making Sex*, but both have also been criticized for the degree of currency they ascribe to the 'one-sex' model of sexual difference in the early modern period; more recent discussions are cited in the endnote.[9] Scholars working on race, too, have explored the way early modern notions of racial difference were informed by conceptual frameworks ranging from the humoral to the religious; two touchstone works on this topic in Shakespeare studies are Kim F. Hall's *Things of Darkness* and Ania Loomba's *Shakespeare, Race, and Colonialism*.[10] And the experience and understanding of the body are also central to other forms of enquiry that have recently begun to inform Shakespeare criticism, such as disability studies and trans studies.[11] The broad term 'embodiment' offers a means of theorizing and critiquing the lived experience and discursive creation of early modern bodies, and of bringing together multiple critical approaches; the 2016 Oxford Handbook on the topic edited by Valerie Traub exemplifies the broad range of analysis that it can enable.[12] The above examples and the accompanying endnotes are very much intended to be illustrative rather than exhaustive, and omit some significant areas of interest (such as the body as focus for anxieties about knowing, and the embodied experience of playgoing) that will be referred to below.

Acknowledging the impossibility of doing justice to this diversity of approaches, the current chapter instead attempts to achieve more limited objectives. To begin with, it explains how a set of ideas identifiable in classical antiquity, subsequently developed in the Islamic world and in medieval Europe, informs Shakespeare's treatment of the body and its workings. Although these ideas have been scientifically discredited, their terminology persists today in ways that can obscure the different implications of the same word in modern and early modern contexts. This chapter also takes account, however, of some ways in which these ideas were being challenged in the early modern period, for example in the anatomical discoveries epitomized in the writings of Vesalius, in ways that, again, discernibly influence Shakespeare's writing. As well as a broad overview, it offers detailed discussion of a small number of Shakespearean texts, beginning with the long poem *Lucrece*, a work whose reliance on narration permits a sustained exploration of the relationship between mind, body, and soul. It then moves on to the *Henry IV* plays, in which Hal's character arc is depicted in humoral terms, and *Othello*, which treats humoralism in relation to racial difference. Finally, this chapter takes account of ways in which the theatre as an institution was affected by early modern beliefs and practices relating to the body and sickness, in particular its vulnerability to legislation intended to limit the spread of plague and the structural affinities between the playhouse and the anatomy theatre.

Galenic Medicine and *Lucrece*

A few lines before identifying the 'Very hand of heaven' in the King's recovery, Lafeu in *All's Well that Ends Well* notes that the patient had been 'relinquish'd of the artists', to which Paroles rejoins 'Both of Galen and Paracelsus' (2.3.10–11). In doing so he juxtaposes two contrasting medical authorities: Paracelsus, the iconoclastic sixteenth-century proponent of novel theories of disease and its cure (discussed in Chapter 2), and the Greek physician Claudius Aelius Galenus (c.130–c.200/c.216 CE), a dominant figure in European medicine for well over a millennium. In his voluminous body of writing, Galen systematized, theorized, and adapted the existing medical learning associated with Hippocrates of Cos, a more shadowy figure alleged to have lived in the fourth and fifth centuries BCE. This tradition largely rejected 'magical or demonic causes of diseases' in favour of a model that is still evident in Shakespeare's works, one key principle of which was an emphasis on the balance within the body of the four qualities of heat, cold, dryness, and humidity.[13] Different combinations of these qualities were manifested in four liquids or 'humours': blood (hot and wet), phlegm (cold and wet), yellow bile (hot and dry), and black bile (cold and dry). (In Shakespeare, the alternative terms 'rheum', 'choler', and 'melancholy' are often used for the last three of these.)[14] Although all four fluids had their basis in actual, observable substances, the Hippocratic–Galenic tradition ascribed them a much more far-reaching significance in the composition of the body and the maintenance of health: for instance, choler served to lighten the blood and clear the body's passageways, while blood carried the other humours around the body, as well as being the material out of which both milk and semen were believed to be refined. Disease was generally understood as 'a loss of … balance' between the humours, whether due to internal or external factors, and this notion that the sick body is one that is out of balance or 'temperature' informs Shakespeare's frequent use of words like 'distempered' to describe it.[15] In *2 Henry IV*, for example, the Earl of Warwick responds to the King's lament that 'the body of our kingdom' has grown 'foul' and diseased by insisting that it is 'but as a body yet distempered; | Which to his former strength may be restored | With good advice and little medicine' (3.1.37–38, 40–42). The prioritization of 'advice' above 'medicine' is significant: according to this therapeutic model, making (or even better, keeping) the body healthy depended in large part on eating the right foods, taking the right kind of exercise, and getting rid of superfluities, whether through natural processes or artificial ones such as purging and bloodletting.

Humoral balance did not merely determine health or sickness: it could also affect an individual's mental disposition. Galen explains in his treatise *The Soul's Traits Depend on Bodily Temperament* that 'when yellow bile abounds in the brain, we are drawn to delirium', and 'when black bile abounds, to melancholia', although he admits to being unsure of the precise causes of these phenomena (subsequent physicians would develop these ideas into a comprehensive theory of temperaments, as will be discussed later on).[16] This represents just one way in which early medicine did not straightforwardly differentiate between what we might call the physical and the mental—the confident assertion of René Descartes in 1641 that 'the mind or soul of man is entirely different from the body' articulates a later separation of these categories that was quite alien to Galenic conceptions of the human.[17] Galen, by contrast, located the rational soul in the brain and understood it as communicating with the rest of the body by so-called 'spirits', 'spirit' being understood as 'a subtle kind of vaporous substance' taken in from the air via the lungs, carried to the heart via the pulmonary vein, and there transformed into life-sustaining 'vital spirit' borne about the body via the arteries. The vital spirits entered the brain via 'a network of fine veins known as *rete mirabile*', where they were transformed into the 'animal spirits' that carried sense perceptions to the rational soul and instructions from it to the rest of the body.[18] 'Soul', incidentally, needs here to be understood as the reasoning faculty rather than the immortal essence assumed by, for example, Christian theology—a discrepancy taken up later in this chapter.

One consequence of Galen's notion of spirits for the reading of Shakespeare is that the term 'spirit' does not necessarily imply a bodiless entity like the ghost of Hamlet's father, or more abstract qualities such as wit or well-being. Instead, it can refer to a material substance with specific biological functions: as the seventeenth-century physician Helkiah Crooke put it, 'A subtle and thin body always moveable, engendered of blood and vapour, and the vehicle or carriage of the Faculties of the soul.'[19] This is what Cornelius has in mind in *Cymbeline* when he explains that the supposed poisons he has given the wicked queen will really only produce the 'show of death', 'the locking up the spirits a time' (1.5.40–41), or Biron in *Love's Labour's Lost* when he laments that 'universal plodding poisons up | The nimble spirits in the arteries' (Additional Passage A, 10–11). The biological meaning sometimes seems to lurk in the background even when other senses are dominant. When Lady Macbeth summons the 'spirits | That tend on mortal thoughts' to 'unsex me here, | And fill me from the crown to the toe top-full | Of direst cruelty' (1.5.39–42), she seems to have in mind some kind of supernatural being; but given that Galen's animal spirits did indeed 'tend on mortal thoughts', and the fact that Lady Macbeth is trying to will bodily change into being ('Make thick my blood', 'take my milk for gall', 1.5.42, 47), a biological interpretation may also be subtly implied.

The image of a thinking soul localized in the brain and engaging with the material world by means of rarefied spirits should not be taken to imply a sovereign intelligence freely directing the actions of the body, however. For one thing, as has already been noted, the soul was itself affected by the fluctuating balance of the humours and their associated qualities. It also had to deal with the 'passions'—love, hatred, joy, grief, and the like—which were understood in terms of a complex relationship between senses, soul, and heart. As Erin Sullivan explains,

sensory information aroused the vital spirits, which communicated with the soul to awaken particular passions. As a passion began to develop, it created motions in the body that drew blood away from one part and towards another, often causing alterations in heartbeat, body temperature, and physical appearance.[20]

Not only did passions disturb the soul: the physical nature of passionate response was such that extreme or prolonged passion could lead to illness or death. Finally, the soul had to deal with the physical desires that Galen located in the third 'governing organ' (after the brain or *encephalon* and heart), the liver:

For as Plato says, the liver is like a wild animal, but this integral part of ourselves must be nourished if there is to be a human race. The reasoning part of us, which is the real man, is situated in the encephalon and has as its servant the irascible [i.e., rational soul] to protect it against this wild animal.[21]

In Galenic medicine, then, the struggle between reason and desire is not something that takes place within a disembodied psyche, nor is it conceived as a battle between mind and body. Rather, it is an ongoing embodied process.

The notion of mental and emotional states as bodily phenomena that Shakespeare's culture inherited from Galenic medicine is both exemplified and problematized in his early narrative poem, *Lucrece*. This is, fundamentally, a text about the relationship between mind and body. At its climax Lucrece, having told her husband and the Roman lords of her rape by Tarquin, asks how she can be cleansed of 'this forcèd stain' and whether her 'pure mind' may 'with the foul act dispense' (1701, 1704). Despite her auditors' insistence that 'Her body's stain her mind untainted clears' (1710), she goes on to take her own life, oppressed both by her sense of herself as example to future women and by her soul's desire to escape 'that polluted prison where it breathed' (1726). The question of whether the body's condition influences that of the soul is, therefore, crucial to the way the heroine's predicament is imagined. However, the physical aspect of mental states is graphically explored much earlier in the poem as it depicts the growth of Tarquin's transgressive sexual desire. This is allotted a specific anatomical location—it is 'the coal which in his liver glows' (47)—and the poem devotes a good deal of space to his attempts to restrain it, which are depicted as an inward 'disputation | 'Tween frozen conscience and hot-burning will' (246–247). As Tarquin proceeds from his bed to Lucrece's chamber, his 'heedful fear …

almost choked by unresisted lust' (281–282), verbal deliberation gives way to mental visions of Lucrece and her husband:

> Within his thought her heavenly image sits,
> And in the self-same seat sits Collatine.
> That eye which looks on her confounds his
>   wits,
> That eye which him beholds, as more divine,
> Unto a view so false will not incline,
> But with a pure appeal seeks to the heart ...
>   (288–293)

Although the notion of the 'mind's eye' is a familiar metaphor, Shakespeare's description here needs to be understood literally. One contemporary text that can help to clarify this is *L'Academie Française*, by the French polymath Pierre de La Primaudaye, the second book of which was printed in English translation in the same year as *Lucrece*. In his discussion of 'the creation, matter, composition, form, nature, profit, and use of all the parts of the frame of man', La Primaudaye includes an account of '*Imagination*, or the *Imaginative* virtue, which is in the soul as the eye in the body, by beholding to receive the images that are offered unto it by the outward senses'. The imagination manages and modifies sense impressions before committing them to memory, a faculty that is specifically localized 'in the hindermost part of the brain, to the end that, after such things as are to be committed unto it, have passed by all the other senses, they should be committed to it to keep, as to their secretary' (see Figure 3.1). This part of the brain is harder and dryer than the rest, partly because it is the beginning of the spinal cord, but also so that memories can be printed there as if by a seal in wax. 'For if it be too soft, the images will be soon ingraven, but they will not stay there any long time ... Contrariwise, if it be over hard, it will be a harder matter to imprint them therein. But when it is well tempered, it receiveth the images easily, and keepeth them well.'[22] Images like those of Lucrece and Collatine in Tarquin's 'thought' could therefore be understood as modified sense-impressions printed in a specific part of the brain.

Tarquin's mind, confronted with the image of Collatine, is admonished of his duty and tries to sway his heart—another instance where the ongoing currency of anatomical terminology as a figure of speech

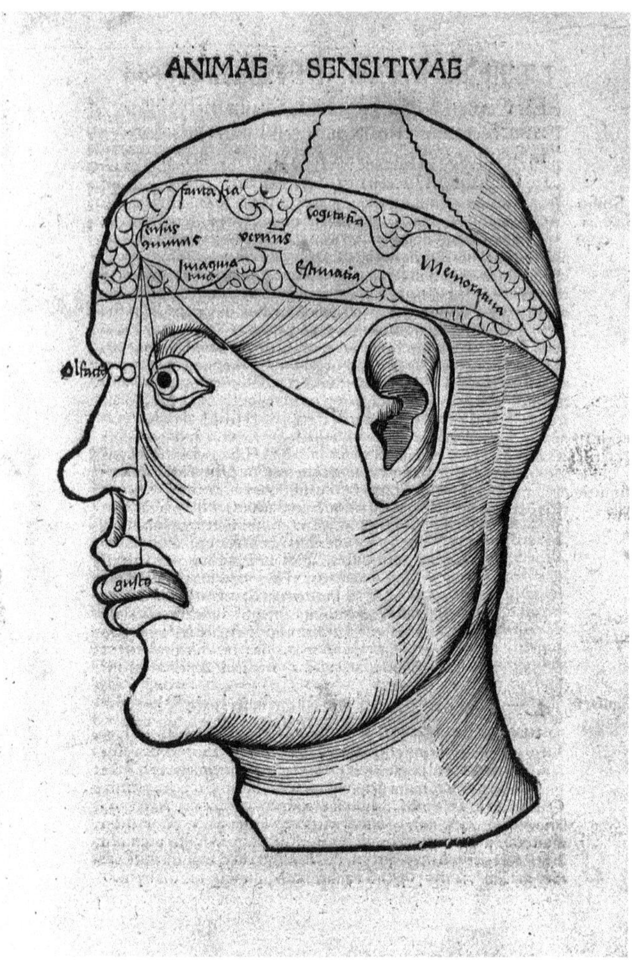

Figure 3.1. Image showing the position of the mental faculties including the imaginative (*imaginativa*, bottom left) and memorative (*memorativa*, right). Gregor Reisch, *Margarita philosophica* (Freiburg, 1503). Wellcome Collection.

for modern readers can obscure its literal significance. Early modern orthodoxy located in the heart 'the power of the Will and Affections', which La Primaudaye defines as 'the motions and acts of that natural power of the soul, which consisteth in following after good, & eschewing of evil'.[23] As such, when Shakespeare writes that Tarquin's heart 'once corrupted takes the worser part', this process can be understood in physical as well as figurative terms. The lines that follow exploit this ambiguity, describing a response that can be read either as emotional or as obscenely physical as his heart 'therein heartens up his servile powers | Who, flattered by their leader's jocund show, | Stuff up his lust as minutes fill up hours' (295–297).

Repeatedly, in the first part of the poem, the heart's role in affective response is presented as an actual biological phenomenon: as Catherine Belling notes, Tarquin's inner disorder 'is positioned explicitly in his blood vessels'.[24] When, on viewing the sleeping Lucrece, Tarquin's eye tempts his veins to 'Swell in their pride', 'his beating heart … Gives the hot charge, and bids them do their liking' (432–434). And when Lucrece expresses the hope that her 'tears', 'sighs', and 'groans' will 'Beat at thy rocky and wreck-threat'ning heart | To soften it with their continual motion' (588, 590–591), her reference to the softening of Tarquin's heart can be taken literally: La Primaudaye explains that while 'they that have a soft and tender heart, receive more easily the impression of joy and grief', 'they that have hard and cold hearts, receive sorrow and grief very soon, and retain it long, as appeareth in melancholy and melancholic persons'.[25]

Shakespeare's depiction of Tarquin in *Lucrece* stresses the embodied nature of psychological and emotional experience: he describes the sight of Lucrece in so far as it affects Tarquin's blood vessels, and his attempt to restrain his lust as a struggle between liver, brain, and heart. Lucrece, however, is driven by the experience of rape to desire to escape from corporeality altogether. Against the possibility that suicide might bring about 'my poor soul's pollution' (1157) she sets the fear that violation of her body will corrupt her soul: 'Ay me, the bark peeled from the lofty pine | His leaves will wither and his sap decay; | So must my soul, her bark being peeled away' (1167–1169). From one perspective, this could be seen as asserting the entwined nature of body and soul: an injury to one injures the other. From another, however, it implies the opposite: the body may house the soul, but the soul is not limited by it, as Lucrece indicates when she says she wants to 'convey this troubled soul' out of 'this blemished fort' (1175–1176). Notwithstanding the poem's ancient Roman setting,

Lucrece's understanding of the soul resembles that of the Christian La Primaudaye in so far as it assumes that souls 'can live well enough, and preserve themselves in their substance, having life always in them, even after they are separated from their bodies'.[26]

Medical Traditions in the Early Modern Period

The way in which *Lucrece* combines an exploration of the bodily dimension of mental experience with an understanding of the soul as something that transcends the body demonstrates the limitations of seeing Shakespeare's medical assumptions purely in terms of the Galenic tradition. For one thing, that tradition had itself undergone significant modification over the centuries. Since the medieval period, Europeans' knowledge of Galen had been mediated through the writings of physicians from the Islamic world, where his works were adopted following the conquest of parts of the Byzantine Empire even as they were largely forgotten in Christian Europe. They were substantially translated in the ninth century by Hunayn ibn Ishaq in Abbasid Baghdad and significantly developed over the years that followed—pre-eminently, by the Persian Ibn Sina (980–1037). His *Al-Qanun fi al-Tibb* or *Canon of Medicine* synthesized Greek medical knowledge with that of the Islamic world, along with the fruits of Ibn Sina's own experience, into a compendious and 'meticulously arranged' account of anatomy, diagnosis, and cure that 'remained the core of the medical curriculum' in Europe until the end of the seventeenth century.[27] When the Galenic tradition made its way back into Europe via southern Italy from the eleventh century onwards, it was in the form of Latin translations not only of Arabic versions of Galen, but also of works by Hunayn (known in Latin as Johannitius), Ibn Sina (Avicenna), and others. It also had to be accommodated to a Christian view of disease that, while not opposed to the practice of medicine, saw sickness as 'a consequence of the Fall of Man, and hence as a consequence of sin' and believed that 'the cure of the soul should take precedence over the cure of the body'.[28] This attitude long remained prevalent: Andrew Boorde's popular *Breviary of Health* (1547), even as it emphasizes the capacity of medicine to extend life, advises 'every sick man … above all things to pacify himself, or to arm himself with patience, and to fix his heart and mind in Christ's death

& passion.... And then if the patient will have any counsel in physic. First let him call to him his spiritual physician which is his ghostly father.'[29] (The Catholic-sounding final phrase is retained through the three reprints in Shakespeare's lifetime.)

Boorde's *Breviary* exemplifies a second reason why an account of Shakespeare and medicine needs to go beyond Galen's writings, namely the fact that Shakespeare's medical reading is much more likely to have focused on relatively contemporary texts printed in the vernacular than on Galen's works or their Latin translations. In some instances critics have identified specific books as likely influences on specific plays—Timothy Bright's *A Treatise of Melancholy* (1586) on *Hamlet*, for example.[30] But beyond these there are a wide array of texts that made medical knowledge available to a lay audience and on which Shakespeare was able to draw. Popular texts such as *The Calendar of Shepherds* (already discussed in Chapter 1) combined their astronomical, astrological, and moral guidance with medical topics such as 'The names of the bones in a man's body', signs of sickness and humoral imbalance, and the foods and activities thought to be good or bad for specific body parts, as well as the zodiacal signs understood to govern them.[31] However, other works offered more detailed accounts of human anatomy and physiology. A good example is Thomas Elyot's *The Castle of Health* (1539), which was, like Boorde's work, regularly reprinted through Shakespeare's lifetime. This text provides an English overview of key concepts in Galenic medicine for educated, well-off male readers: its title advertises a book 'whereby every man may know the state of his own body, the preservation of health, and how to instruct well his physician in sickness, that he be not deceived'. The key concepts include 'Things Natural' such as the four elements, their possible combinations in different 'complexions', the four humours, the members or parts of the body, etc.; 'Things not natural', namely 'Air', 'Meat and drink', 'Sleep and watch', 'Moving and rest', 'Emptiness and repletion', and 'Affections of the mind'; and 'Things against Nature', in effect 'pathological conditions of all kinds'.[32] The 'diligent consideration' of these 'three sorts of things' (referred to by modern medical historians as naturals, non-naturals, and contra-naturals) helps to keep the human body healthy. Accordingly, Elyot goes into great detail about the importance of avoiding corrupted air; the effects of different foods and drinks; how much time should be devoted to sleep and exercise; how to remedy the 'superfluous abundance of humours in the body'; and how to govern the 'affects and passions of the mind', which 'if they be immoderate, they do not only annoy the body, & shorten the life, but also they do appair, and sometime lose utterly a man's estimation'.[33]

A complicating factor in this approach to maintaining health is that humans are naturally predisposed to certain different 'complexions', or combinations of heat, cold, dryness, and moisture. For example, 'The Body, where heat and moisture have sovereignty, is called *Sanguine*, wherein the Air hath pre-eminence, and it is perceived and known by these signs, which do follow'—Elyot's list here includes 'veins and arteries large', 'The visage white and ruddy', 'Dreams of bloody things, or things pleasant', 'Angry shortly', and 'The urine red and gross'.[34] The phlegmatic was typically more fat, slow, and of weak digestion; the choleric lean, hardy, and sharp-witted; and the melancholic fretful, insomniac, and slow to digest. This meant that materials or practices beneficial to one temperament could be harmful to another: 'hot wines, pepper, garlic, onions, and salt, be noyful [harmful] to them, which be choleric, because they be in the highest degree of heat and drythe, above the just temperance of man's body in that complexion. And yet be they oftentimes wholesome to them, which be phlegmatic.'[35]

While the idea that different individuals had different complexions or temperaments went back to Galen, a significant innovation of Arab physicians, developed by their followers in Christian Europe, was the association of these with 'psychological or emotional characteristics'.[36] By Shakespeare's time this was a commonplace, and it frequently provides him with a vocabulary with which to describe both transient moods and more deep-seated personality traits such as the 'melancholy disposition' of Don John in *Much Ado About Nothing* (2.1). In *The Taming of the Shrew*, Petruccio uses it as his justification for withholding from Katherina mutton that he says is overcooked:

> I tell thee, Kate, 'twas burnt and dried away,
> And I expressly am forbid to touch it,
> For it engenders choler, planteth anger,
> And better 'twere that both of us did fast,

> Since of ourselves ourselves are choleric,
> Than feed it with such overroasted flesh.
>   (4.1.156–161)

Petruccio's insistence that he and Katherina, being choleric, need to avoid the burnt meat is arguably in line with the observation of Levinus Lemnius that 'hot, fat and sweet meats … are very apt to be turned into Choler'.[37] On the other hand, Elyot writes that because choler is an aid to digestion, 'in a choleric stomach, by abstinence, these inconveniences do happen, humours adust, consuming of natural moisture, fumosities and stinking vapours, ascending up to the head', as when a kettle boils dry.[38] From this perspective, Petruccio's starving of Katherina is medically illiterate as well as inhumane.

In the canon of early modern drama, the term 'humour' is inextricably linked with the 1590s comedies of George Chapman (*The Blind Beggar of Alexandria* of 1596, *An Humorous Day's Mirth* of 1597) and Ben Jonson (*Every Man in His Humour* (1598), *Every Man Out of His Humour* (1599)). These plays, however, are only tangentially informed by the underlying medical model. As Asper explains in the Induction to *Every Man Out*:

> in every human body,
> The choler, melancholy, phlegm, and blood
>   …
> Receive the name of humours. Now, thus far
> It may by metaphor apply itself
> Unto the general disposition,
> As when some one peculiar quality
> Doth so possess a man that it doth draw
> All his affects, his spirits, and his powers,
> In their confluxions all to run one way.
> This may be truly said to be a humour. (Induction, 96–97, 100–107)[39]

The term 'humour' is here used metaphorically to mean a 'peculiar quality' that distorts an individual's behaviour, generating the play's gallery of eccentrics and obsessives such as the travel-addicted 'vain-glorious knight' Puntarvolo and the 'affecting courtier' Fastidious Brisk (Characters, 11, 28). However, the behaviour of Jonson's characters is further modified by their own awareness of humours theory, which leads them to new extremes of delusion and pretentiousness. In *Every Man In*, the 'country gull' Stephano and the urban poetaster Matheo both lay claim to the fashionable humour of melancholy, supposedly associated with poetic genius:

> STEPHANO   Ay, truly, sir, I am mightily given to
>   melancholy.
> MATHEO   Oh, Lord, sir, it's your only best humour, sir.
>   Your true melancholy breeds your perfect fine wit, sir.
>   I am melancholy my self divers times, sir, and then
>   do I no more but take your pen and paper presently,
>   and write you your half-score or your dozen of
>   sonnets at a sitting. (Quarto version, 2.3.64–68)

Both kinds of character—the eccentric and the affected—have their Shakespearean equivalents: in the Henriad and *The Merry Wives of Windsor*, for example, we encounter the ridiculous Pistol, whose speech is confected from the cliches of 1580s drama, and the roistering Nim, who endlessly rationalizes human behaviour in humoral terms ('I will cut thy throat ... that is the humour of it', *Henry V*, 2.1.67–68). However, at other points in his career Shakespeare can also be seen employing humoral theory in a way that, while it can be metaphorical or satirical, sometimes seems intended more literally. The varied use of it in *Romeo and Juliet* is a case in point. Outside the Capulets' orchard, Mercutio tries to summon the absent Romeo with the words 'Romeo! Humours! Madman! Passion! Lover!' (2.1.7), apparently using 'humours' as a byword for the abnormal behaviour assumed to accompany erotic desire. At other times, 'humour' can denote a damp or liquid entity (not necessarily one of the four bodily humours), as when Benvolio speculates that Romeo wants to consort with 'the humorous night' (2.1.31) or when Friar Laurence tells Juliet that after taking his potion 'through all thy veins shall run | A cold and drowsy humour' (4.1.95–96). But the play's references to humours can sometimes be more specific. Chapter 1 has already noted how the violence that pervades Verona in *Romeo and Juliet* is explained in terms of hot weather aggravating choler, a word that appears in the play's third line of dialogue; as Benvolio later puts it, 'now, these hot days, is the mad blood stirring' (3.1.4). And when Montague says of his son's behaviour 'Black and portentous must this humour prove, | Unless good counsel may the cause remove' (1.1.138–139), 'humour' can be understood both as an abstract term denoting Romeo's mood, and as a more literal reference to the black bile whose predominance may have caused it. Similarly, when Romeo, climbing into the Capulets' orchard, exclaims 'Turn back, dull earth, and find thy centre out' (2.1.2), he uses the Aristotelian conception of earth being irresistibly drawn to the centre of the cosmos as a metaphor for his attraction to Juliet; but the image gains a further level of complexity from the identification of melancholy with the element of earth in humoral theory. La Primaudaye, for example, writes that 'the muddy dregs, which commonly thicken & settle in the bottom of [the blood], are of the nature of the earth, & are called Melancholy'.[40] Romeo's association of himself with the earth may signal that the predominance of melancholy his father discerned in him persists, despite the transfer of his affections from Rosalind to Juliet.

The Humours in *Henry IV*

While the influence of humoral physiology can be discerned in numerous Shakespearean texts, it is arguably in the two *Henry IV* plays that it is most pervasive, insofar as the changing balance of humours can be seen as shaping the plays' overall trajectory. At the end of the first scene in which we encounter him, Prince Hal is already promising to 'throw off' the 'loose behaviour' (1.2.205) he manifests in his drinking and thievery with Falstaff/Oldcastle and his other Eastcheap companions, and he fulfils this promise at the end of Part 2 when he banishes Falstaff with the words 'I know thee not, old man. Fall to thy prayers' (5.5.47). Critics have conceptualized this character arc in various ways, from E. M. W. Tillyard's comparison of Hal to the protagonist of a Morality Play to Stephen Greenblatt's view of him as deliberately enacting a temporary subversion in order to repudiate it when he becomes king.[41] However, the prolonged exercise in self-control (including the appearance of loss of control) that Hal undertakes can also be seen in terms of humoral management, a reading that Hal himself invites when he promises to 'a while uphold | The unyoked humour of your idleness' (1.2.192–193) before undergoing a future 'reformation' (1.2.210). As Michael Schoenfeldt points out, the way humoral physiology links mental states to humoral balance, and humoral balance to non-naturals such as diet, exercise, and repletion, invites 'the consuming subject … to conceive all acts of ingestion and excretion as very literal acts of self-fashioning'.[42] William Vaughan's *Natural and Artificial Directions for Health*, for example, itemizes the effects upon different bodily complexions of different foodstuffs:

> Young hen Partridges eaten with vinegar do heal all manner of fluxes, and dry up bad humours in the belly….
> Quails eaten with coriander seed and vinegar do help melancholic men….
> Woodcocks and Snipes are somewhat lightly digested. Yet hurtful for choleric and melancholic men….
> Weightier Oranges are very good for them that be melancholic, and keep back the rheum….
> Plums and damsons do qualify blood, and repress choleric humours.[43]

Falstaff parodies ideas like these in Part Two when he contrasts Hal with the 'sober-blooded' Prince John: 'Hereof comes it that Prince Harry is valiant; for the cold blood he did naturally inherit of his father he hath, like lean, sterile, and bare land, manured, husbanded, and tilled, with excellent endeavour of drinking good, and good store of fertile sherry, that he is become very hot and valiant' (4.2.13–18). However, his ironical testimony to the Prince's dissoluteness is truer than he knows, for it is just this kind of deliberate, self-administered treatment that has led Hal to associate with Falstaff and his other companions while fully intending to drop them when it becomes expedient. As Hal puts it after his tormenting of Francis the drawer— himself something of a humours character in the Jonsonian sense, unable to say much beyond 'Anon, anon, sir' and rendered immobile when summoned by two customers at once—'I am now of all humours that have showed themselves humours since the old days of goodman Adam to the pupil age of this present twelve o'clock at midnight' (2.5.93–96). The implication seems to be that Hal's time in Eastcheap gives him access to a diversity of experience that equips him for future kingship.

Drinking aside, Hal's time with Falstaff and the others is largely only analogous to humoral management, rather than involving the actual ingestion of substances that will produce a change in humoral balance. However, the *Henry IV* plays set him up in contrast to other characters whose lack of control is repeatedly depicted in humoral terms. On the one hand there is Hotspur, 'altogether governed by humours' (3.1.230) and variously presented as 'drunk with choler' (1.3.127) and as given over 'To thick-eyed musing and curst melancholy' (2.4.46). As Lady Percy tells him, 'A weasel hath not such a deal of spleen | As you are tossed with' (2.4.76–77), using a synonym for melancholy drawn from the organ in which it was believed to reside. In being 'governed by a spleen' (5.2.19), Hotspur displays 'want of government' (180), both of himself and, by extension, of others; this positions him as a foil to Hal, whose claim to self-mastery validates him as a future ruler. Hal's other foil in this respect is Falstaff, whom he hyperbolically characterizes as 'that trunk of humours, that bolting-hutch of beastliness, that swollen parcel of dropsies, that huge bombard of sack, that stuffed cloak-bag of guts, that roasted Manningtree ox with the pudding in his belly' (2.5.454–458). The grotesque language he uses to describe Falstaff links overindulgence in food and drink with both physical size and humoral chaos: his body is not an orderly structure but a distended receptacle for 'humours' and 'dropsies' or watery swellings. In contrast to the 'unruly' Falstaff, whose physicality is constantly emphasized, Hal is, as David Hillman puts it, 'almost disembodied, a lean and hungry cipher who tightly controls his words, his body, his surroundings … he "knows" his companions' "unyok'd humors" only in order to reject them, to "purge" (3.2.20) himself and his eventual kingdom'.[44] As the observer or manipulator in others of humours he is able to control in himself, Hal anticipates Jonsonian characters like Lorenzo Junior in *Every Man In*, who enjoys playing off the 'humour' of the rustic Stephano (2.3.49) against the urban Matheo and who ends up married to Hesperida at the expense of less intelligent and self-aware characters.

Othello, Humours, and Race

In the context of *Henry IV*, Hal is a somewhat ambiguous figure who can come across either as a diligent monarch-in-training preparing to draw together a divided kingdom, or as a cold and unappealing manipulator who exploits his friends in the cause of public relations and uses his time with them as a kind of self-administered medicine. Iago, whose conscious villainy in *Othello* is responsible for multiple deaths and great suffering, certainly lacks Hal's amiable qualities and sense of public duty; but in his detached and deliberate attitude to self-fashioning he can at times seem like a nightmare version of the same character:

Virtue? A fig! 'Tis in ourselves that we are thus or thus. Our bodies are our gardens, to the which our wills are gardeners; so that if we will plant nettles or sow lettuce, set hyssop and weed up thyme, supply it with one gender of herbs or distract it with many, either to have it sterile with idleness or manured with industry, why, the power and corrigible authority of this lies in our wills. If the beam of our lives had not one scale of reason to peise another of sensuality, the blood and baseness of our natures would conduct us to most preposterous conclusions. But we have reason to cool our raging motions, our carnal stings, our unbitted lusts; whereof I take this that you call love to be a sect or scion. (1.3.319–332)

Iago's references to nettles, lettuce, hyssop, and thyme appear as part of an extended metaphor that imagines the body as a garden, so perhaps they should not be taken too literally as dietary suggestions. However, all four had their place in humoral medicine: according to Elyot, lettuce 'maketh abundance of blood', 'provoketh sleep', 'increaseth milk in a woman's breasts', and 'abateth carnal appetite', while hyssop 'digesteth slimy phlegm', nettles purge it, and thyme purges melancholy.[45] Indeed, it is hard to disentangle the literal from the metaphorical in Iago's speech: its final sentence recommends reason as a bridle to the passions, but since the preceding sentence locates those passions in the body (as per the Galenic notions already discussed in relation to *Lucrece*), it may be that Iago really is recommending that Rodrigo cool his 'carnal stings' and 'unbitted lusts' by eating more lettuce. It is notable that the verb 'manure', which Iago uses to describe the will's action on the body, is one that also appears in Falstaff's account of Prince Hal, who has manured 'the cold blood he did naturally inherit of his father' by drinking copious sack: Iago recalls Hal in conceiving of self-control in terms that are both metaphorically and literally humoral. This is reinforced by the imagery of purging that occasionally appears in his speeches, as when he says to Rodrigo of Othello and Desdemona that 'The food that to him now is as luscious as locusts shall be to him shortly as bitter as coloquintida' (whose inner fruit 'hath the nature to purge', according to William Turner's *A New Herbal* (1551; see *OED* 'coloquintida, *n.*')), and when he wishes that the fingers that Cassio keeps putting to his lips 'were clyster-pipes [enema tubes] for your sake' (1.3.47–49, 2.1.179–180). Paster notes 'the centrality to early modern medical practice of the alimentary purge, administered in the form of emetics, laxatives, and enemas, which was experienced from time to time by nearly everyone from infancy on', and proceeds to speculate on the psychosexual consequences of this particular form of bodily regulation.[46] In light of this it might be questioned whether Iago's sense of the body as something to be governed and managed reflects unusual will-power and self-mastery, or instead represents something more pathological and obsessive.

While Iago represents a rather strained form of self-control, insistently focused on the lower bodily strata, a different form of the same impulse is evident in the play's title character. A defining gesture in the play's first act is his authoritative quelling of potential violence between his supporters and Brabantio's ('Keep up your bright swords, for the dew will rust 'em', 1.2.60), and the episode seems to externalize the impulse to self-government that he later expresses when insisting to the senate that he is asking for Desdemona to follow him to Cyprus only 'to be free and bounteous to her mind', not 'To please the palate of my appetite' (1.3.265, 262). Responding to the disorderly acts of Cassio—'rash and very sudden in choler', according to Iago (2.1.271)—Othello provides a commentary on his affective response that simultaneously advertises the mastery involved in controlling that response: 'Now, by heaven, | My blood begins my safer guides to rule, | And passion, having my best judgement collied, | Essays to lead the way' (2.3.197–200). 'Collied' primarily means 'blackened' (as in 'the collied night' in *A Midsummer Night's Dream*, 1.1.145), and as Paster suggests, the imagery can be taken literally as an account of the bodily nature of passionate experience: 'his heart's blood is heating in anger, and the sooty excrements formed in that central bodily furnace ascend to the brain to darken the rational faculties situated there'.[47] At the same time, in punning on the 'choler' to which Cassio has fallen prey the word also sets up a contrast between the abilities of Othello and of his lieutenant to subdue their anger.

However, in the context of a play where references to blackness, while always racially coded, call attention to their racial significance to an unusual degree, Othello's description of his judgement as 'collied' by passion inevitably raises the question of how the language of humoral theory and of the passions intersected with early modern understandings of racial difference. This chapter has already mentioned Montague's description of Romeo's melancholy as a 'black' humour, and historians of early modern racialization have noted how some writers on humoralism linked warm climates to 'an abundance of black bile'.[48] Desdemona, however, discussing with Emilia Othello's possible response to her loss of the handkerchief he gave her, responds to the enquiry 'Is he not jealous?' with the observation, 'Who, he? I think the sun where he was born | Drew all such humours from him' (3.4.29–31). In view of later developments in the play, this statement is clearly ironic, and it seems to contradict some assumptions about the relationship between climate and jealousy that were prevalent in the early modern period. As Ania Loomba notes, 'Robert Burton's *The Anatomy of Melancholy* suggests that "Southern men are more hot, lascivious, and jealous, than such as live in the North"', and while Burton's text postdates *Othello*, it draws on writers such as Jean Bodin and Leo Africanus whose works would have been available to Shakespeare.[49] However, humoral theory also licensed contrasting ways of thinking about the relationship between climate and temperament: Mary Floyd-Wilson cites authorities such as pseudo-Aristotle and the medieval philosopher Albertus Magnus who ventured opinions comparable to Desdemona's. Albertus writes of Ethiopians, for example, 'because their bodies are surrounded by very hot air, it is necessary that they be porous and dry, since the moisture evaporates continually from them'. This loss of moisture produces a greater 'refinement of their spirits … they excel more in ingenuity on account of the moving heat'. It is also imagined as being responsible for colour difference: 'their bodies grow dark on account of the scorching of the body; for it gives off a fine moisture and it burns the earthly mass which remains, and generates blackness'.[50]

The examples given by Floyd-Wilson do not diminish the very real anti-black racism that informs *Othello*, but they do suggest how 'malleable' humoral theories of racial difference could be, variously positioning Africans as hot-blooded and intemperate or as 'sober because they are cool'.[51] Furthermore, as Kim F. Hall points out, to explain such difference in terms of the action of the sun on the body and its humours was also to assume an underlying similarity. She cites the Prince of Morocco's speech to Portia in *The Merchant of Venice*, 'Mislike me not for my complexion, | The shadowed livery of the burnished sun' (2.1.1–2): although it produces bodily difference, the potential for sunburn is something that 'connects white European and black African', and thus 'becomes an important way of negotiating difference through a shared sameness'.[52] And while Morocco seems to intend 'complexion' primarily in the sense in which it is understood today, as denoting skin colour, this chapter has already noted the more general meaning of humoral balance that was still current in Shakespeare's time. As such, in identifying himself in terms of his complexion Morocco is also alluding to a sense of selfhood as residing in the humours that he shares with Portia, different though the proportions of their humours may be.

However, a theory of racial difference that emphasized the role of the climate had also to account for the fact that blackness was evidently transmissible to one's descendants even if one moved to a more northerly clime and had a white sexual partner—something that Iago exploits when warning Brabantio that 'the devil will make a grand-sire of you' (1.1.91), and that is visibly realized in *Titus Andronicus* in the form of the black child that Tamora bears Aaron. Humoral theorists did argue that acquired characteristics could be passed down through generations: Juan Huarte, for example, notes that 'though 200 years have passed over our heads, since the first Egyptians came out of Egypt into Spain, yet their posterity have not forlorn that their delicacy of wit and promptness, nor yet that roasted colour which their ancestors brought with them from Egypt'.[53] However, another frequently cited explanation is that volunteered by George Best in his account of the voyages of exploration he undertook with Martin Frobisher in the 1570s:

I myself have seen an *Ethiopian* as black as a coal brought into *England*, who taking a fair English woman to Wife, begat a Son in all respects as black as the Father was, although *England* were his native Country, & an English woman his Mother: whereby it seemeth this blackness proceedeth rather of some natural infection of that man, which was so strong, that neither the nature of the Clime, neither the good complexion of the Mother concurring, could anything alter, and therefore we cannot impute it to the nature of the Clime.[54]

Best's account is shaped both by racist assumptions and by an outlook that to the modern reader appears 'scientific': he deduces from the fact that an African fathers a black child to a white mother in a northern country that blackness cannot derive from 'the nature of the clime'. However, his further explanation of the 'natural infection' rests not on a consideration of observable phenomena, but on an apocryphal religious tradition: he ascribes it to the transgression of Noah's son Ham, who disobeyed his father's commandment not to have sex while in the ark.[55] In consequence, 'God would a son should be born, whose name was Chus [Cush], who not only itself, but all his posterity after him, should be so black & loathsome, that it might remain a spectacle of disobedience to all the World.' Best's linkage of blackness with deviant sexuality participates in a discourse whose influence can be discerned in numerous early modern plays, not least *Othello*; but it also serves to exemplify how fields of enquiry that to the modern reader might seem scientific (namely, in modern terms, genetics) were not straightforwardly separable from religious belief. Furthermore, discussions like those of Best and Huarte testify to the central role of race in early modern speculations about reproductive biology. As a physical characteristic that was observably passed to one's offspring, skin colour was an obvious topic for consideration by writers interested in the workings of heredity, and this means that the history of this branch of medicine is also a history of racial thinking: as Loomba asserts, 'Theories of conception, of generation, of transmission … cannot be isolated from anxieties about racial difference'.[56]

Anatomy and Ocular Proof

So far this chapter has considered *Othello* as a play informed by humoral theory, including the way this conception of the body dealt with the racialized topic of skin colour. However, it is also a play that is deeply and actively concerned with the wider problem of what humans can know about the body, and by what means, a question that is most obviously thematized in Othello's desire for certainty as to whether Desdemona has committed adultery with Cassio. When Othello demands of Iago, 'Villain, be sure thou prove my love a whore. | … Give me the ocular proof' (3.3.364–365), Iago is able to supply three pieces of evidence: his presumably fictional account of Cassio talking in his sleep in 3.3.415–430; his (again fictional) account of Cassio wiping his beard with the handkerchief Othello gave Desdemona (3.3.437–444), corroborated in Othello's mind by Desdemona's actual loss of the handkerchief; and his conversation with Cassio in 4.1, which Othello sees, but cannot hear, during which Cassio exhibits gestures that Iago has primed Othello to think refer to Desdemona but actually relate to Bianca. Othello's behaviour here can be seen as an example of the tendency of the 'green-eyed monster' jealousy (3.3.170) to colour the viewer's perception, and as such it exemplifies a central problem of experimental science: how to interpret external phenomena in a way that is not affected by the prejudices of the observer.[57] However, Iago describes Othello's predicament not as a problem in the viewer, but as a problem inherent in the phenomena that he wants to view:

> Would you, the supervisor, grossly gape on?
> Behold her topped?…
> It were a tedious difficulty, I think,
> To bring them to that prospect….
> It is impossible you should see this,
> Were they as prime as goats, as hot as
>   monkeys,
> As salt as wolves in pride, and fools as gross
> As ignorance made drunk. But yet I say,
> If imputation, and strong circumstances
> Which lead directly to the door of truth,
> Will give you satisfaction, you might ha't.
> (3.3.400–403, 407–413)

Iago's argument relies on sleight of hand: it starts by asserting that it would be 'tedious' to arrange a situation in which Desdemona and

Cassio might be caught in adultery and then shifts the terms, stating that it would be not just difficult but 'impossible'. This alleged impossibility is used by Iago to justify conjecture (informed by 'imputation, and strong circumstances') as being the nearest thing available to the 'ocular proof' Othello has demanded, and in the process the difference between proof and conjecture slips out of view.

Iago's sophistry can be read in characterological terms, as an expression of his cynicism and capacity for manipulation. However, it can also be seen as having implications that go beyond its immediate dramatic function and that raise more fundamental questions relating to the body as an object of enquiry. While he is wrong to state that the prospect of catching two lovers in the act is theoretically impossible rather than merely difficult, that does not dispel the difficulty of knowing the truth about something more genuinely inaccessible, namely the question of Desdemona's chastity—as Iago later puts it, 'Her honour is an essence that's not seen' (4.1.16). In the wider play, this abstract notion has a material stand-in in the form of Desdemona's lost 'handkerchief | Spotted with strawberries' (3.3.439–440), its design suggestive of hymeneal blood. In Iago's speech in 3.3, it is perhaps more elliptically implied by his comment that 'imputation, and strong circumstances' can 'lead directly to the door of truth': the truth of Desdemona's chastity is imagined in spatial terms as residing within a building or behind a wall, protected by a door that may or may not be penetrable. The possibility, or the impossibility, of obtaining knowledge through access to the body's physical interior has already been raised by Iago earlier in the scene when he responds to Othello's line 'By heaven, I'll know thy thoughts' with the words 'You cannot, if my heart were in your hand; | Nor shall not whilst 'tis in my custody' (3.3.166–168). The body is here depicted both as something that might be opened up in search of truth, and as something that might stubbornly refuse to yield to such efforts.[58]

The image of Othello holding Iago's heart in his hand, like an executioner or anatomist, exemplifies a trope of seeking truth through bodily dismemberment that can be found elsewhere in Shakespeare. In *King Lear*, for instance, Edgar justifies opening the dead Oswald's correspondence on the grounds that 'To know our enemies' minds we'd rip their hearts; | Their papers is more lawful' (*History of King*

Lear, 4.5.252–253), while in the Folio version Lear during the mock-trial urges, 'let them anatomize Regan; see what breeds about her heart. Is there any cause in nature that makes these hard-hearts?' (*Tragedy of King Lear*, 3.4.34–36). The judicial setting here is appropriate, given the overlap between punitive and medical contexts for opening the body in the early modern period: one the one hand, Catholic priests and others found guilty of treason were executed by means of what was in effect vivisection, while on the other, a specified number of corpses of condemned criminals were granted to London's barber surgeons and physicians to be dissected.[59] Speeches like these, not to mention the more literal scene of bodily violation that is staged in Gloucester's blinding, blur the boundary between the Shakespearean playhouse and its near namesake, the anatomy theatre. In so doing, they make reference to a way of knowing the body that was enabling significant challenges to existing medical orthodoxy.

By the early modern period, the dissection of corpses had long been a feature of European medicine. It had begun 'by the end of the thirteenth century', according to Nancy Siraisi, at universities such as Bologna, Padua, and Montpellier, albeit on a small scale due to the limited availability of corpses and religious and cultural taboos.[60] A select number of medieval anatomists were therefore able to do something that Galen could not: although human dissection was practiced for a time in Alexandria during the third century BCE, it was generally forbidden in the ancient Grecian and Roman worlds, and Galen's extensive dissections were constrained by a 'necessary reliance on animals'.[61] However, direct observation of internal human anatomy seems to have been used in the late Middle Ages as a means of illustrating Galen's teachings rather than generating challenges to any of the errors that arose from this reliance.[62] Well into the sixteenth century the Parisian anatomist Jacobus Sylvius, teacher of Vesalius (see below), held so confidently to Galen's authority 'that he insisted that any dissected corpse exhibiting a deviation from the master's teaching must be a negligible exception'.[63]

By Shakespeare's time, however, not only were anatomists beginning to chart the inner workings of the body with increasing accuracy; as they did so, they were able to point to errors in the writings of Galen, Ibn Sina, and other authors, and they could also draw on the growing reach and sophistication of print technology to publicize their findings. An epoch-making work in this respect was *De humani corporis fabrica* ('On the Structure of the Human Body') by the Brussels-born Andreas Vesalius (Andries van Wezel, 1514–1564), first printed in 1543 and then revised in 1555, whose account of the structure of the human skeleton, muscles, blood vessels, nervous system, and organs was accompanied by lavish, sometimes grimly beautiful engravings. Though far from being the first illustrated anatomical work to be printed, it was unprecedented in the extent to which both its text and its images were the result of direct observation of human cadavers; Vesalius avowedly joins Galen in exhorting medical students towards 'dissections carried out with their own hands'. However, while celebrating Galen as 'easily among the foremost professors of dissection', Vesalius also notes that 'Galen deviated much more than two hundred times from a true description of the harmony, use, and function of the human parts', and his description rejects numerous features of Galenic anatomy including the *rete mirabile* (which does not, in fact, exist in humans). Against established authority he sets up the value of seeing for oneself either the cadaver in the flesh or (the next best thing) his book's illustrations, which 'will place the dissected body before the eyes of those studying the works of Nature'.[64]

The engravings accompanying *De fabrica*, frequently plagiarized over the decades that followed its publication, offered one means by which those who were not medical professionals could obtain 'ocular proof' of the body's inner workings—albeit at second hand. Another was depicted in stylized form on its frontispiece, which represents a scene of dissection in what is functionally, if not architecturally, an anatomy theatre. As Jonathan Sawday explains, this was a type of structure that fascinated English travellers to cities such as Bologna, Leiden, and Padua, whose public dissections could not be matched by 'the occasional demonstrations given at the College of Physicians and the Hall of the Barber-Surgeons in London'. However, Vesalius's title page (see Figure 3.2) demands to be read not as an accurate depiction of one of these spaces, but as a 'drama of life and death' in which living onlookers mingle with classical figures and death's skeletal personification. Sawday's account underscores how its subject is a theatre not just in the etymological sense of being a place for viewing things, but in the multiple scenes it enacts: the escaping

Figure 3.2. Title page of Andreas Vesalius, *De humani corporis fabrica* (Basel, 1543). National Library of Medicine, USA.

monkey, the men who 'squabble over surgical instruments' beneath the table, the importunate figure with the dog. At its centre lies the female corpse whose womb the anatomist, Vesalius, reveals to the gaze of seated onlookers and reader alike.[65] Sawday describes anatomy theatres as 'the playhouses of organized violence' whose performances 'rivalled the stage', and this is clearly a comparison that works both ways: critics have readily noted scenes in early modern drama where bodies 'are *opened*, like bodies in anatomy theatres'.[66] Along with the blinding in *Lear*, David Hillman cites the tongue-biting in Thomas Kyd's *The Spanish Tragedy* and the heart-ripping in John Ford's *'Tis Pity she's a Whore*, but more extreme examples might be adduced: George Peele's *The Battle of Alcazar* includes a dumb-show where a sheep's heart, lungs, and liver are apparently used to represent the internal organs of human characters, while the protagonist of Henry Chettle's *Hoffman* exhibits the anatomized corpse of his father while taking his bloody revenge.[67] At moments like these the distinction between playhouse and anatomy theatre can seem, not just to blur, but to collapse altogether.

King Lear and *Titus Andronicus* aside, Shakespeare's spectacles of dismemberment tend to be less gruesome than these; but one instance of metaphorical dissection that seems especially suggestive in light of the Vesalius frontispiece relates to the scene in which Hamlet intrudes into his mother's closet (a symbolically charged setting if ever there was one). Responding in soliloquy to Gertrude's summons after the performance of the *Mousetrap*, he addresses his heart:

> Let not ever
> The soul of Nero enter this firm bosom.
> Let me be cruel, not unnatural.
> I will speak daggers to her, but use none.
> (3.2.382–385)

In stating his reluctance to follow Nero's example, Hamlet invokes an emperor who not only was responsible for the death of his mother Agrippina, but also (in a spurious legend to which Shakespeare alludes in *King John*, 5.2.152) was believed to have 'caused his mother's womb to be ripped (she living) to see the place where he lay in'.[68] When Hamlet later confronts Gertrude in her closet, the audience's memory of this line adds an ironic appropriateness to her response to his injunction, 'You go not till I set you up a glass | Where you may see the inmost part of you': 'What wilt thou do? Thou wilt not murder me?' (3.4.19–21). She seems to take Hamlet's remark as a statement of intention to literally show her her own insides; and while this is, strictly speaking, a misinterpretation, his earlier allusion to the archetypal matricidal anatomist suggests that she has somehow discerned a genuine, if repressed, impulse on his part. Her cry for help prompts Hamlet to a different but analogous act of violence: not the opening of her body to reveal its internal organs but a thrust into the arras to reveal Polonius, whose corpse Hamlet will later dismissively refer to as 'the guts' (3.4.186). Verbally in the reference to Nero, and symbolically in the killing of Polonius, *Hamlet* gestures towards a revelation of the kind depicted on the front of *De fabrica*, namely the anatomization of the female body to unveil the mystery of its generative parts. Hillman, in a nuanced and detailed discussion of this scene, refers to the 'urge to open the parental body—to know one's parent's entrails' in the context of a wider discussion of Hamlet's apparent view of truth as 'an internal matter, to be discovered by getting inside something', and this chimes with the way *Othello* uses the bodily interior as a way of addressing themes of knowledge and doubt.[69] Beyond Hamlet's own desires, however, the scene can also be considered in relation to the theatrical spectacle it seems to threaten, or offer: a public anatomical demonstration akin to those available on the Continent or, less celebratedly, at the Barber-Surgeons' Hall.

Although the Shakespearean stage was, as has already been noted, frequently capable of staging scenes of violence and dismemberment, the kind of anatomization to which Hamlet alludes was clearly beyond it. It was not, however, beyond the narrative poem, and the example of Hamlet and Gertrude offers a productive basis from which to review a text already considered in this chapter, namely *Lucrece*. The passages quoted earlier from that poem represented her death as a means by which her soul might be liberated from various forms of oppression; but the moment when she is said to have 'sheathèd in her harmless breast | A harmful knife' (1723–1724) also produces a bodily revelation manifested in her blood:

> And bubbling from her breast it doth divide
> In two slow rivers, that the crimson blood
> Circles her body in on every side ...
> Some of her blood still pure and red
> remained,

> And some looked black, and that false
>   Tarquin-stained.
>
> About the mourning and congealèd face
> Of that black blood a wat'ry rigol goes,
> Which seems to weep upon the tainted place;
> And ever since, as pitying Lucrece' woes,
> Corrupted blood some watery token shows;
>
> And blood untainted still doth red
>   abide,
> Blushing at that which is so putrefied.
>   (1737–1739, 1742–1750)

The blood shed by Lucrece's corpse, dividing into differently coloured streams, is understood as revealing the truth both of her account of Tarquin's rape and of her own claim of 'untainted' innocence; as such, while Lucrece falls short of exposing her internal organs, it offers the kind of bodily confirmation sought for by Othello and Hamlet alike. In making it manifest to a group of aristocratic male spectators, Lucrece defers to the social group that Steven Shapin identifies as crucial in verifying observational data, as was noted in Chapter 1. At the same time, in turning her own body into an anatomical demonstration, she arrogates to herself the role of anatomist so authoritatively bestowed on Vesalius on the frontispiece of *De fabrica*. Like Helen in *All's Well*, she transgresses into a way of acting on the body that is usually deemed to be a masculine preserve; but where the spectacle of Helen's pregnant body at her play's climax still leaves space for Bertram's 'If', Lucrece's act of self-inflicted violence suggests what the logical end point of male desire for knowledge over female sexuality may be.

The Body of the Audience

This chapter has outlined some ways in which Shakespeare's writing is informed by medical ideas prevalent in his time, in particular about the relationship between mind and body and about the role of the humours in both health and subjectivity. It has also shown how, in his repeated use of imagery suggestive of anatomization, Shakespeare registers a way of knowing the body that had already thrown up challenges to Galenic orthodoxy and would continue to do so over future decades, William Harvey's theory of circulation of the blood being an obvious example. The following brief conclusion touches on the question of how the theatre, as an institution, was itself implicated in the medical practices and beliefs of Shakespeare's day.

One of these has been noted already, namely the structural analogy between the early modern playhouse and the anatomy theatre that dramas such as *Hamlet* seem to exploit. In offering up, or at least alluding to, spectacles of bodily dismemberment, playing companies placed their audiences in positions comparable to those of the onlooking throng on the frontispiece of *De fabrica*. But the varying responses expressed in the bodies of Vesalius's spectators—rapt, horrified, indifferent, distracted—reveal the extent to which playgoing was, itself, an embodied experience. As Matthew Steggle points out, 'contemporary accounts of and allusions to early modern theatre are fascinated by audience laughing and weeping, which is considered an important element within the performance occasion'.[70] The physician and essayist Sir Thomas Browne wrote in 1642 that he could 'weep most seriously at a Play, and receive with a true passion, the counterfeit griefs of those known and professed impostures', while a decade earlier the antitheatrical writer William Prynne asked, 'what other pretence have Play-haunters for their resort to Play-houses … but to pass away the time in mirth? to laugh till their sides do ache again, at the Clown's behaviour, or some other merry jests and passages'.[71] These tangible physical responses were seen as consequence of the power of drama to stimulate the passions in a process that—as in the case of Tarquin in *Lucrece*—was understood very much in bodily terms. This could be a positive thing: in the induction to *The Taming of the Shrew*, Sly is encouraged to watch a play on the grounds that

> your doctors hold it very meet,
> Seeing too much sadness hath congealed your
> blood,
> And melancholy is the nurse of frenzy.
> Therefore they thought it good you hear a play
> And frame your mind to mirth and
> merriment,
> Which bars a thousand harms and lengthens
> life. (Induction 2, 127–132)

However, it also had the potential to make playgoing, as Alison P. Hobgood points out, 'an intensely corporeal, highly emotive activity characterized by risky, even outright dangerous bodily transformation'.[72] To this extent, a productive, if anachronistic way of thinking about early modern plays might be as experiments conducted by dramatists and playing companies on the bodies of their audiences: will this scene make playgoers laugh? Will that one make them weep?

Passionate response was not the only form of bodily risk to which early modern playgoers were exposing themselves. Theatres were well understood to be environments favourable to the spread of disease, specifically plague, and they were accordingly shut down on a regular basis when pestilence was prevalent or feared. In April 1580 the Privy Council forbade 'all plays within and about the City until Michaelmas [i.e., 29 September] next for avoiding of infection', a pattern regularly followed in summer for most of the 1580s, and the theatres were closed due to plague for most of the period between summer 1592 and summer 1594 (not to mention lengthy closures in the 1600s).[73] Although this was no doubt a sensible measure, the simultaneous insistence on keeping churches open reveals the tangle of religious and epidemiological imperatives underlying it. As the preacher Thomas White put it in a sermon of 1577 that welcomed the closing of theatres, 'the cause of plagues is sin, if you look to it well: and the cause of sin are plays: therefore the cause of plagues are plays'.[74] In White's reading, playhouses are a health hazard not because they bring infected people into proximity with one another but because they incur the wrath of God, who punishes London with pestilence accordingly.

Furthermore, the seeming obviousness to the modern mind of theatre closures as a measure against the plague (especially given recurrent lockdowns during the ongoing Covid pandemic) can obscure the extent of its inconsistency with medical orthodoxy. In Galenic theory, disease derives from imbalance of the humours: in Elyot's words, 'by the increase or diminution of any of them in quantity or quality, over or under their natural assignment, inequal temperature cometh into the body, which sickness followeth'.[75] As Jonathan Gil Harris notes, this physiological model was 'often at a loss to account for [the] transmission' of contagious diseases since 'it regarded disease not as an ontological thing that invades the body, but as a state of imbalance within it', and its proponents were obliged to fall back on explanations reliant on the influence of non-naturals such as corrupt air upon the humours.[76] The notion of disease as an actual external phenomenon was associated with radical thinkers such as Paracelsus, who believed 'that each disease had an ontological cause: an "archeus" or spirit that had entered into the body and set itself against the local archei necessary for the proper functioning of the body', and the Veronese physician Girolamo Fracastoro, who gave the disease syphilis its name.[77] In his Latin poem on the subject, Fracastoro offered the atomist suggestion (see Chapter 2) that 'the seeds of so great a pestilence … are in the very air, which circulates itself through every land and insinuates itself throughout our bodies', from which it can be transmitted to others.[78]

However, orders for the closure of theatres make no reference to any such theories, or indeed to the Galenic orthodoxy that they challenged. Instead, they presumably drew on practical experience of the fact that social mixing was somehow conducive to 'infection', irrespective of how infection itself was understood to take place. To this extent, the very working of the early modern theatre was constrained by medical thinking, even if that medical thinking struggled to find a home for itself within the prevalent orthodoxy of its day.

| 4

Maths

Earlier chapters of this book have highlighted moments in Shakespeare's drama when the playhouse seems analogous to spaces of scientific enquiry such as laboratories, observatories, or anatomy theatres. A different kind of comparison is offered at the beginning of *Henry V*, when audiences are explicitly invited to see the playhouse as something more akin to a multiplication device:

> But pardon, gentles all,
> The flat unraisèd spirits that hath dared
> On this unworthy scaffold to bring forth
> So great an object. Can this cock-pit hold
> The vasty fields of France? Or may we cram
> Within this wooden O the very casques
> That did affright the air at Agincourt?
> O pardon: since a crookèd figure may
> Attest in little place a million,
> And let us, ciphers to this great account,
> On your imaginary forces work. (Prologue, 8–18)

Requesting pardon for the disparity between the grandeur of the historical victory that will be represented and the unworthiness of the actors and their theatre, the Chorus reaches for a mathematical analogy. Just as a numerical symbol such as '1' can, if followed by six other digits, signify not the number 'one' but 'one million', so the Chorus asks the audience to treat individual actors as signifying greater numbers, whose magnitude is indicated a few lines later with the line, 'Into a thousand parts divide one man' (24).[1] This is a notion

Shakespeare and Science. Tom Rutter, Oxford University Press. © Tom Rutter (2024).
DOI: 10.1093/oso/9780192898548.003.0005

that the Chorus continues to develop in line 17, when the actors are described as 'ciphers to this great account'. Although by 1599 the word 'cipher' had already acquired its more familiar modern-day meaning of a secret code, its original referent in English—one more relevant to the point the Chorus is making here—was the arithmetical symbol that we now call zero, 'cipher' coming via multiple European languages from the Arabic *ṣifr*, which is in turn 'simply a translation of the Sanskrit name *śūnya*, literally "empty"' (*OED* 'cipher', *n.*). As the symbol 'o' had been introduced to Christian Europe at the same time as the other Hindu-Arabic numerals 1–9, 'cipher' could in different contexts mean either 'o' specifically or, in more general terms, any of those ten numerals. The Chorus's lines thus imply both the relative littleness of the Shakespearean actors and their power to denote larger things: if the record of Henry's famous victories is imagined, according to a financial metaphor, as a 'great account', then the actors are mere digits by comparison, but just as a digit (depending on its position in a string of numerals) can indicate a number of units, or of thousands, or of millions (see Figure 4.1), so an actor playing one soldier, indulged by the audience's imagination, can be taken to represent multitudes. This effect is enhanced by the Chorus's description of the playhouse as a 'wooden O', alluding to the circular shape suggested by the many-sided building. The ambivalent zero has the power both to represent nothingness and, when placed to the right of another digit, to multiply the number that digit signifies by ten; similarly, the zero-shaped playhouse may be 'unworthy' in itself, but when combined with the 'crooked figure' of the actor it is able to magnify that figure from one man to a whole battalion.

Hindu-Arabic numerals were not a new arrival in sixteenth-century England: Shakespeare's compatriots had been aware of them for centuries, as I will discuss in more detail below. However, they had come into widespread use only recently and Roman numerals were still employed by many, as were old-fashioned methods of calculation such as the use of counters, perhaps explaining the sense of strangeness that can be discerned in the words of the Chorus: as Eugene Ostashevsky notes, 'Hindu-Arabic notation must have seemed so novel that people still could wonder at it.'[2] There is assumed to be something uncanny about the way the value of a figure can change depending on

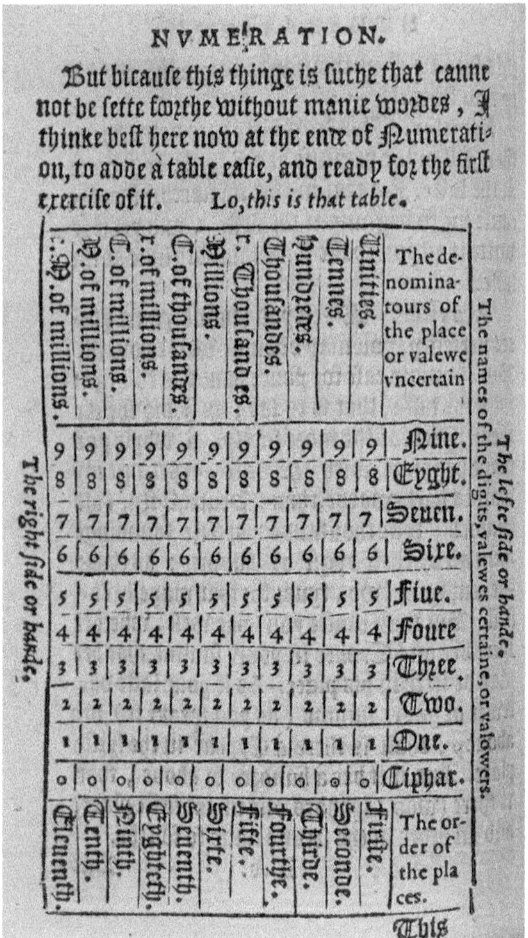

NVME RATION.

But bicause this thinge is suche that canne not be sette foorthe without manie woordes, I thinke best here now at the ende of Numerati‑ on, to adde a table easie, and ready for the first exercise of it.   Lo, this is that table.

The denominatours of the place or valewe vncertain	M. of millions.	M. of millions.	C. of millions.	r. of millions.	Millions.	C. of thousandes.	r. of thousandes.	Thousandes.	Hundreds.	Tennes.	Unities.
Nine.	9	9	9	9	9	9	9	9	9	9	9
Eyght.	8	8	8	8	8	8	8	8	8	8	8
Seuen.	7	7	7	7	7	7	7	7	7	7	7
Sixe.	6	6	6	6	6	6	6	6	6	6	6
Fiue.	5	5	5	5	5	5	5	5	5	5	5
Foure	4	4	4	4	4	4	4	4	4	4	4
Three.	3	3	3	3	3	3	3	3	3	3	3
Two.	2	2	2	2	2	2	2	2	2	2	2
One.	1	1	1	1	1	1	1	1	1	1	1
Ciphar.	0	0	0	0	0	0	0	0	0	0	0
The order of the places.	Eleuenth.	Tenth.	Ninth.	Eyghteth.	Seuenth.	Sixte.	Fifte.	Fourthe.	Thirde.	Seconde.	Firste.

Figure 4.1. Table showing the different values of Hindu-Arabic digits depending on their position. Robert Recorde, *The ground of arts* (London, 1573). The Rare Book & Manuscript Library, University of Illinois at Urbana-Champaign.

its position, and about the contradictory qualities of the cipher or zero, in a way that is hard for modern minds brought up on Hindu-Arabic numerals to appreciate.

This chapter is about the impact of those numerals on Shakespeare's writing and thought, along with other basic mathematical concepts such as definitions of the circle (in *Henry V*) and the language of division (in *King Lear*). If these themes seem rather limited and prosaic compared with the astronomical ideas discussed in Chapter 1 or the fascination with the body's interior alluded to in Chapter 3, that is because they are: Shakespeare does not seem to have possessed the same level of mathematical understanding as, for example, his contemporary Ben Jonson, who owned a copy of Henry Billingsley's translation of Euclid's *Elements of Geometry* which he demonstrably read, as well as alluding in *The Magnetic Lady* (1632) to the 'logarithms' (1.6.35) described by John Napier in 1614.[3] By contrast, nothing in Shakespeare's writings implies a complex or theoretical appreciation of mathematical concepts beyond what an early modern property owner and company sharer might need. Although this chapter will allude to the importance of mathematics in the history of early modern science, this is one field in which it is particularly appropriate to observe Mary Thomas Crane's recent injunction 'not to go beyond "what was in the air", and to avoid arguing for a technical knowledge that most playwrights and poets did not possess'.[4]

Shakespeare's limited technical knowledge in mathematics did not, however, prevent mathematical concepts from influencing his creative imagination in a deep and sustained way. Paula Blank has identified what she calls 'Shakespeare's rhetoric of measurement', whereby 'dramatic characters and poetic personae regularly speak of numbering, weighing, or measuring one another', and this concern with metrics and value is evident on a larger scale in the titles of works like *Much Ado About Nothing*, *Love's Labour's Lost*, and (of course) *Measure for Measure*.[5] However, as the example of *Henry V* shows, Shakespeare was also intrigued by aspects of mathematics that had less to do with rational calculation and more to do with the strange behaviour, the visual appearance, even (as will be discussed below) the sexual suggestiveness of numbers themselves. That makes his use of mathematics typical of the approach to scientific concepts set out in this book, whereby a deep thematic concern with the possibilities, implications, and limits of early modern ways of knowing the universe sits side by side both with an opportunistic concern for poetic effect and with an openness to subversiveness and play. This multiplicity is well exemplified in *King Lear*, which combines a profound questioning of the feasibility and desirability of quantifying such things as love, land, and life itself with a discourse of foolery that subjects numbers and counting to ridicule: 'The reason why the seven stars are no more than seven is a pretty reason'. 'Because they are not eight'. 'Yes. Thou wouldst make a good fool' (*History of King Lear*, 1.4.34–38).

The paragraphs that follow offer a fuller rationale for the study of mathematics in Shakespeare that makes reference to early modern English attitudes to the subject, advances in pure and applied mathematics in the period, and the importance of mathematics in emergent scientific practice. I then attempt to contextualize Shakespeare's treatment of Hindu-Arabic numerals by explaining how they differed from other methods of representing and working with numbers, noting some reasons for their (gradual) uptake in Europe, and discussing the textbook likeliest to have provided Shakespeare with his introduction to them, Robert Recorde's *Ground of Arts*. There follow discussions of two plays that seem especially preoccupied both with mathematical concepts and with the visual icons with which such concepts were rendered, namely *Henry V* and *King Lear*.

Why Study Shakespeare and Mathematics?

As has already been intimated, extended discussion of Shakespeare's use of mathematical ideas in the prologue to *Henry V* may seem somewhat out of proportion to the fairly basic nature of the processes it describes: multiplying by tens (or multiples of ten) and dividing one by a thousand are operations that modern readers are likely to have learned in primary school. However, as Keith Thomas suggests in his essay 'Numeracy in Early Modern England', while the ability to count was all but universal in Shakespeare's time, levels of facility in more complex mathematical skills, although hard to deduce with certainty, seem to have varied widely.[6] As their name implies, grammar schools prioritized literacy over numeracy: the seventeenth-century schoolmaster John Brinsley devotes hardly any of his *Ludus literarius: or, The Grammar Schoole* to the mathematical curriculum, and has one of his speakers lament that 'you shall have scholars, almost ready to go to the University, who yet can hardly tell you the number of pages, sections, chapters, or other divisions in their books, to find what they should'.[7]

This deficiency seems to have stemmed in part from a perception that mathematics were the business of tradesmen and merchants rather than of scholars and gentlemen. Thomas notes the comment of the braggart Armado in *Love's Labour's Lost* that the skill of 'reckoning' 'fitteth the spirit of a tapster' (1.2.40–41), as well as an observation by the seventeenth-century antiquarian John Aubrey that a 'Bar-boy at an Alehouse will reckon better and readier than a Master of Arts in the University'; while allowance needs to be made for rhetorical exaggeration, such remarks indicate that mathematical ability was prized neither as a social asset nor as an accomplishment within the humanist educational system. Furthermore, arithmetic was generally perceived to be difficult. Edmund Wingate wrote in 1630 that 'most men despair to attain the knowledge thereof', not just because of the obscurity of the treatises that discuss it, but because multiplication and division 'so confound, and perplex the new *Practitioner*, that he takes them to be *Hercules' Pillars*, and writes upon them *Non plus ultra*'—no further.[8]

One justification for including this chapter, then, is the possibility that the number-games played by the Chorus in *Henry V* may have seemed a little less obvious, and a little more recondite, to early play-goers than they now appear. Another, though, is the fact that despite the patchiness of mathematical education the sixteenth and seventeenth centuries were a period of considerable change in this field. The gradual adoption of Hindu-Arabic numerals facilitated new techniques such as 'the introduction of decimals, logarithms and algebra', while the period also saw advances in many varieties of applied mathematics such as navigation, cartography, and surveying.[9] Although Shakespeare's level of familiarity with the former is open to question, his works frequently demonstrate an interest in what his contemporaries sometimes referred to as the 'mechanical arts'.[10] In *Shakespeare and Technology: Dramatizing Early Modern Technological Revolutions*, Adam Max Cohen gives numerous examples of Shakespeare depicting or referring to the fruits and practices of arts such as navigation, cartography, ballistics, and clock-making, sometimes using strikingly technical vocabulary (as with Claudius's repeated deployment in *Hamlet* of terms associated with gunnery such as 'proof', 4.7.126, and 'blank', 4.1.39). He also points out that as a landowner, Shakespeare may well have been familiar with, or at least benefited from, the considerable advances in the technology of surveying that took place in his era.[11] The way in which early modern drama registered changes both in the practice of mathematics and in the real-world application of mathematical techniques is a topic of growing interest to critics: Joseph Jarrett finds a 'generation' of playwrights 'being inspired by the new cultural prominence that mathematics achieved in the sixteenth and early seventeenth centuries, whereby rapid and radical innovations in the subject began to transform society, and mathematical pedagogy became an increasingly necessary task for a broad spectrum of institutions and individuals'.[12]

Not only were early modern dramatists influenced by developments in the teaching, practice, and application of mathematics; the stage itself constituted one material context in which mathematics were applied. As Henry S. Turner emphasizes in *The English Renaissance Stage: Geometry, Poetics, and the Practical Spatial Arts, 1580–1630*, the theatre was a place where multiple ways of working with space, such as architecture, ballistics, and sculpture, converged; moreover, in an era when 'demography, statistics, economics, sociology, or political science did not exist in their modern disciplinary forms, the "inventions" of early-modern writers provided a method for modelling large-scale social processes', as with the depiction of international conflict in miniature within *Henry V*. According to this way of thinking, theatre not only benefits from 'spatial arts' in which mathematics are applied; it is a form of applied mathematics itself, a means of representation analogous to cartography or perspective drawing. Turner notes the way in which terms drawn from the mechanical arts, such as 'plot', 'engine', and 'device', are deployed in descriptions of theatrical technique—strikingly, in Ben Jonson's *Every Man Out of His Humour*, a play with which *Henry V* is closely contemporaneous.[13]

One final reason why a discussion of mathematics belongs in this book is because of the conceptual and practical centrality of mathematics to the developments in science that were taking place in Shakespeare's era. An important feature of the 'new science' practised by Galileo and others was the attempt to view the universe quantitatively: as Gary Hatfield puts it, 'the doctrine that the material world consists of bodies having only mathematical properties and that it must therefore be comprehended and described in mathematical terms'.[14] The book of the universe, Galileo wrote in 1623, cannot be understood unless one first learns to comprehend the language and interpret the characters in which it is written. It is written in the language of mathematics, and its characters are triangles, circles, and other figures, without which it is humanly impossible to understand a single word of it; without these, one is wandering about in a dark labyrinth.[15]

This belief can be seen as a defining quality of emerging scientific thought in the early modern period. Paula Blank writes that 'It is a tenet of the history of science that the modern disciplines of astronomy, medicine, music, navigation, physics and other arose when Renaissance natural philosophers replaced the "qualitative" view of reality dominant among the ancients with a "quantitative" view based, centrally, on procedures of measurement.'[16] The distinction can be overstated: mathematics did have a central place in some branches of medieval learning, of which astronomy—with its emphasis on predicting the movement of heavenly bodies through accurate measurement and the use of geometry—is perhaps the most obvious. And although the medieval university curriculum enshrined the *trivium* of grammar, logic, and rhetoric at undergraduate level, Master's students went on to study the *quadrivium* of arithmetic, geometry, astronomy, and music (which was understood to be underpinned by mathematical proportions). Nevertheless, when seventeenth-century theorists of science attempted to identify what was new about their approach by comparison with that of the ancients they were apt to reach, like Galileo, for mathematics as offering a precise and objective way both of acquiring and of representing knowledge about the universe. When in his 1667 *History of the Royal Society* Thomas Sprat set out to explain how the method of the 'new Undertakers' who had founded the Society was 'different from that, on which the former have proceeded', a key tenet was the need 'to separate the knowledge of *Nature*, from the colours of *Rhetoric*, the devices of *Fancy*, or the delightful deceit of *Fables*'. As he explained later in his history, members were instead expected to employ 'a close, naked, natural way of speaking; positive expressions; clear senses; a native easiness: bringing all things as near the Mathematical plainness, as they can'.[17] Sprat's invocation of mathematics here neatly exemplifies Mary Poovey's contention that in the early modern period numbers were coming to acquire 'connotations of transparency and impartiality' and the reputation of being 'somehow *epistemologically* different from figurative language', 'value-free'.[18] The confident words of Sprat, five decades after Shakespeare's death, position the dramatist's own age as one in which not only the practices and applications, but also the cultural meanings of mathematics were in a state of transition. The paragraphs that follow outline some aspects of that transition, beginning with the one Shakespeare dwells on at such length in *Henry V*: the growing use of Hindu-Arabic numerals, and the new ways of thinking about number that they enabled.

Representing Numbers in Early Modern England

As with other branches of learning considered in this book, many of the fundamental assumptions of early modern mathematics can be traced back to ancient Greece (which was in turn influenced by other cultures including the Egyptians and the Babylonians). One example is the tendency to define a number as a collection of units, so that 'one' is not, properly, a number at all—as Shakespeare puts it in Sonnet 136, 'Among a number one is reckoned none'. This definition stemmed from a conception of mathematics as an intellectual activity, befitting high social status, which involved 'the contemplation of that which is in no way subject to change', namely 'the 'continuous' magnitudes—lines, areas, solids', and 'the "discrete" numbers—two, three, four, etc'.[19] These activities were distinguished from the mundane activities of measuring quantities and of carrying out the recording, adding, and subtracting necessary for buying and selling, and they excluded from consideration concepts such as fractions (as not representing an entity that could be counted) and irrational numbers such as the square root of two. This idea of mathematics as 'a search for larger truths', as Michele Jaffe puts it, also created an intellectual climate that was inhospitable to the concept of zero: since dominant conceptions of the universe did not allow for the existence of void space, there was no such thing as nothing.[20] Greek mathematicians were familiar with the idea of zero as used by the Babylonians when carrying out astronomical calculations, but avoided using it in their own numerical notations.[21]

When representing numbers in writing, the Greeks used the letters of their alphabet to signify 1–9, 10 and its multiples to 90, and

100 and its multiples to 900.[22] Because different letters were used to signify (for example) 9, 90, and 900, a letter's numerical value did not depend on its position relative to other letters: the symbols for 5, 40, and 300 would amount to 345 irrespective of the order they were placed in. The Roman system for representing numbers—still widely used in Shakespeare's day—was very different, but like that of the Greeks it ascribed an inherent value to numerical symbols irrespective of place order. At its most basic level the Roman system uses a stroke to signify one, and further strokes to signify further ones: thus, while in the Hindu-Arabic '11' the numeral on the right stands for one and the one on the left stands for ten, in the Roman 'II' each numeral stands for one, producing a total of two. Higher numbers are signified by V (five), X (ten), L (fifty), C (a hundred), D (five hundred), and M (a thousand). While the convention has arisen that placing a smaller number before a larger one signifies a degree of diminution (so that the number four can be written both as IIII and as IV), this was not always the case: according to Jaffe, ICI, IIC, and CII were all accepted ways of representing the number 102, and the influence of Hindu-Arabic numerals may have been a factor in the growth of standardization in Roman from the sixteenth century onward.[23]

While 'the convention that concatenation was to be interpreted as addition' gives Roman numerals a certain referential simplicity, they are cumbersome for handling large numbers: compare 1999 with MDCCCCLXXXXVIIII, for example, and think about how you might go about doing sums with the latter.[24] In practice, the Romans—like many other cultures—solved this problem in day-to-day arithmetic by using counters, the abacus, and similar instruments rather than written numerals (see Figure 4.2). However, their numerical system made it tedious both to make more complex calculations and to represent them in writing. By contrast, the Hindu-Arabic system represents large numbers economically by giving the numerals 1–9 a fluctuating value depending on their position in a longer string where the first number from the right represents units, the second tens, the third hundreds, and so on. However, in order to do this unambiguously—for example, to show that the '1' furthest from the right in '1001' means a thousand and not ten—it requires a numeral to show that there are (for example) no tens and no hundreds, thereby distinguishing 1001 from 11.

Figure 4.2. The Greek mathematician Pythagoras calculates with counters while the sixth-century Roman Boethius uses Hindu-Arabic numerals, with which some later writers anachronistically believed him to have been familiar. Gregor Reisch, *Margarita philosophica* (Freiburg, 1503). Houghton Library, Harvard University.

Recognizable versions of the numerals 1–9 were in use in India by the fifth century CE, and the *śūnya* or zero by the ninth.[25] Following Muslim conquests in India in the eighth century, first by the Umayyads and then by the Abbasids, they came into use in Islamic cultures, most notably in the great intellectual centre of Baghdad that also produced Arabic translations of Aristotle (see Chapter 3). They were incorporated within foundational works of mathematics such as Muhammad ibn Musa Al-Khwarizmi's treatise on algebra, the *al-jabr wa l'muqabala*, as well as enabling the advances on Greek astronomy discussed in Chapter 1.[26] As with other scientific texts, trade, conflict, and other forms of contact between Christian and Islamic worlds brought works like these into Christian Europe: the *al-jabr wa l'muqabala* was translated into Latin by Robert of Chester in the twelfth century, while Al-Khwarizmi's textbook on arithmetic using Hindu-Arabic numerals, translated by Adelard of Bath, was widely adapted by other authors such as John of Seville, Johannes de Sacrobosco, and Alexandre de Villedieu.[27] It was also via Latin versions of Arabic translations that Greek texts such as Euclid's *Elements of Geometry* became available.

Notwithstanding the evident familiarity of learned Europeans from the twelfth century onward with Hindu-Arabic numerals, it was several centuries before they became firmly established, for reasons that are not altogether clear. Although historians of the zero such as Charles Seife and Robert Kaplan cite the persistent suspicion, going back to the Greeks, of concepts of the void and of nothingness, the visual unfamiliarity of the symbols, their counterintuitive behaviour, and their origins in non-Christian cultures may also have been off-putting.[28] A medieval English translation of Villedieu's poem *De algorismo*, accompanied by a prose commentary, explains its title (actually derived from the name of Al-Khwarizmi, also retained in our 'algorithm') with the words 'There was a king of Ind, the which heyth [was called] Algor, & he made this craft'. It explains that when using the 'ten signs of Ind', 'thou shall write backward', that is, from right to left, starting with units on the right-hand side. As for the signs themselves, 'Every of these figures betokens himself & no more, if he stand in the first place of the rule…. If it stand in the second place of the rule, he betokens ten times himself.' However, 'A cipher tokens naught, but he makes the figure to betoken that comes after him more than he should & [=if] he were away'.[29] Indian signs written from right to left, which mean different things depending on their position, and another sign that means nothing but changes the meaning of other signs: it's easy to imagine that this unfamiliar way of using written symbols could have seemed more like magic than a reputable branch of scholarship.

Notwithstanding their strangeness, the usefulness of 'signs of Ind' in carrying out complex calculations ultimately led to their widespread adoption. Merchants and bankers, in particular, found them invaluable, since they facilitated processes such as currency conversions and working out interest. Moreover, the new technique of double-entry book-keeping that was developed in fourteenth-century Italy prioritized the balancing of debit and credit accounts, and hence the accurate recording and tallying of lengthy columns of numbers.[30] As a sharer in the Lord Chamberlain's (subsequently King's) Men, a part-owner of the Globe playhouse, and a substantial property-owner, Shakespeare would have been well aware of accounting practices, and references to them appear throughout his work well beyond the description of *Henry V* as 'this great account'—an epithet that puns on the word's associations both with storytelling and with finance. The two preceding plays in the second tetralogy of Histories both include substantial extended metaphors drawn from accounting, the first when Prince Hal, taken to task for his prodigality, promises his father that he will one day make the admired Hotspur 'exchange | His glorious deeds for my indignities':

> Percy is but my factor, good my lord,
> To engross up glorious deeds on my behalf;
> And I will call him to so strict account
> That he shall render every glory up,
> Yea, even the slightest worship of his time,
> Or I will tear the reckoning from his heart. (3.2.145–152)

Here, Hotspur is imagined as a Hal's financial agent, acquiring honours that he must ultimately 'render ... up' to the Prince as if undergoing an especially rigorous audit. And in the play's sequel, Morton responds to Northumberland's rage and grief at Hotspur's death by reminding him that he must have allowed for the possibility of it when encouraging him into battle: 'You cast th' event of war, my noble lord, | And summed the account of chance, before you said | "Let us make head"' (1.1.165–167).

New Economic Criticism, as represented by works including Linda Woodbridge's edited collection on *Money and the Age of Shakespeare* and Peter Grav's *Shakespeare and the Economic Imperative*, has emphasized the pervasiveness of such habits of thought within Shakespeare's writing, and as Natasha Korda notes in an essay on the Sonnets and *The Merchant of Venice*, 'The language of (ac)counting in these and other dramatic texts is richly suggestive of the central role played by bookkeeping and formal instruments of credit, such as bonds, in the expanding credit economy that gave rise to the commercial theaters.'[31] Take, for example, Portia's speech in *The Merchant of Venice* in which she sets out what Bassanio has gained in choosing the right casket:

> You see me, Lord Bassanio, where I stand,
> Such as I am. Though for myself alone
> I would not be ambitious in my wish
> To wish myself much better, yet for you
> I would be trebled twenty times myself,
> A thousand times more fair, ten thousand times more
>   rich,
> That only to stand high in your account
> I might in virtues, beauties, livings, friends,
> Exceed account. But the full sum of me
> Is sum of something ... (3.2.149–158)

These lines well exemplify Korda's contention that Portia's language throughout the play is 'replete with references to accounts, full sums, terms in gross, oaths of credit, sureties, and the like'.[32] While presenting herself, with ostensible humility, 'Such as I am'—'an unlessoned girl, unschooled, unpractisèd' (159)—she also expresses the wish that her desirable attributes might be multiplied so as to give her a higher valuation in Bassanio's 'account'. In monetizing herself in this way she seems to turn herself into an object that, like her dowry, will pass into Bassanio's possession with marriage; but in attaching a condition to the transfer of property Portia makes it not a gift but a loan akin to the one Shylock offers Antonio:

> This house, these servants, and this same myself
> Are yours, my lord's. I give them with this ring,

> Which when you part from, lose, or give away,
> Let it presage the ruin of your love,
> And be my vantage to exclaim on you. (170–174)

Korda links the adroitness with which Portia manages herself and her property to the role played by female moneylenders and theatre shareholders such as Susan Baskerville in the development of the early modern theatre industry; the financial themes of plays such as *Merchant* are hard to detach from the commercial environment for which they themselves were written.[33]

Henry V and Recorde's Arithmetic

An important resource that helped equip Elizabethan businessmen such as Shakespeare with the mathematical skills they needed was the array of textbooks that came into print during the sixteenth century. Probably the most widely read maths primer was *The Ground of Arts* (1543), by Robert Recorde, whose *Castle of Knowledge* has already been discussed in Chapter 1. This accessible and wide-ranging textbook had gone through around two dozen printings by 1612, when Brinsley referred to 'Record's Arithmetic' as the standard introduction to the subject, and would see the same number by the end of the century.[34] It starts at the first principles of numeration and the use of Hindu-Arabic numerals and runs through techniques of addition, subtraction, multiplication, division, and the so-called Golden Rule before showing examples of their practical application such as calculating the profits due to different shareholders in an enterprise. A second part, in print by 1552, discusses fractions and introduces other techniques of calculation; a third, not by Recorde, added in 1582 includes 'simple interest, loss and gain, bartering and exchange rates'.[35]

Although Recorde's emphasis on the ways mathematical techniques could be used in real-life contexts such as buying merchandise, paying labourers, and erecting buildings is one feature of his book that might have appealed to Shakespeare, another is its dialogic form, which has a Master teaching 'the art of numbering' to a mathematically incompetent Scholar. This setup allows for the explanation of basic ideas such as place value and the signification of the different Hindu-Arabic numerals, but it also dramatizes the educational process, as when the Scholar gets his figures mixed up:

> MASTER   How write you these four other: ii. i. ix. viii.?
> SCHOLAR   Thus (I trow): 2. 1. 6. 8.
> MASTER   Nay, there you miss: Look on mine example again.
> SCHOLAR   Sir truth it is; I was to blame, I took 6 for 9, but I will be warer hereafter.
> MASTER   Now then take heed, these certain values every figure representeth, when it is alone written without other figures joined to him. And also when it is in the first place, though many other do follow, as for example, This figure 9 is .ix. standing now alone.
> SCHOLAR   How is he alone and standeth in the middle of so many letters?
> MASTER   The letters are none of his fellows. And if you were in France in the middle of M. [=1,000] French men, if there were no English man with you, you would reckon yourself to be alone.[36]

Not only does the dialogue here generate a degree of gentle comedy; referring to the 9 as 'he' rather than 'it', and as a 'figure' that stands in a certain 'place', serves to humanize the unfamiliar Hindu-Arabic numerals in a way that anticipates Shakespeare's Chorus referring to the actors as 'ciphers' and a single digit as a 'crookèd figure'. Brian Rotman emphasizes that Shakespeare was 'in the first generation of children in England to have learned about zero from Robert Recorde's *Arithmetic*', so direct influence is not implausible.[37] Furthermore, the Master's distinction between the Hindu-Arabic 'figures'—1, 2, 3—and the 'certain values' that 'every figure representeth'—one, two, three—treats the former as signs rather than things in a way that makes them highly useful to Shakespeare's metatheatrical discussion of the relationship between individual human actors and the armies and kings that they are taken to stand for. More tentatively, one might point out that the prospect of being 'here alone' and surrounded by Frenchmen is one that the common soldier Bates wishes on King Henry on the eve of Agincourt (*Henry V*, 4.1.120).

Biographical explanations for Shakespeare's interest in maths in *Henry V* are both unnecessary and unprovable, but it may be worth repeating that the play was written at a period of significant professional transition for Shakespeare. By the end of 1598, he was part of a consortium, along with Cuthbert and Richard Burbage, John Heminges, William Kempe, Augustine Phillips, and Thomas Pope, that leased the land in Southwark on which the Globe playhouse was built and would pay for its construction, the Burbages owning 25% shares and the rest 10%, although they 'absorbed Kempe's share when he left the company in 1599, so that for a time each owned 12.5 per cent'.[38] As such, he changed from being merely an actor-sharer in the Lord Chamberlain's Men to being part-owner of a playhouse. *Henry V* touches on the fluctuating economics of share-ownership when Henry makes the best of the inferior English numbers by observing, 'The fewer men, the greater share of honour' (4.3.22), and insisting, 'I would not lose so great an honour | As one man more methinks would share from me | For the best hope I have' (31–33). More obviously, while the play may have appeared before the company started playing at the Globe, the Chorus's opening reference to 'this wooden O' seems at least to anticipate a move to a venue whose name emphasized circularity.

As the opening paragraphs of this chapter demonstrated, the prologue to *Henry V* exploits the resemblance between a (not quite-) circular playhouse and the mathematical figure of the zero or cipher in the context of a discussion of the power of the actor not just to represent another individual, but to stand symbolically for multitudes. However, elsewhere in the play characters build on the imagery established in the prologue in ways that emphasize not the zero's arithmetical properties, but the circle's geometrical ones. Witness, for example, the speeches of the Duke of Exeter and the Archbishop of Canterbury in 1.2 explaining how the King can go to war in France while leaving men in England to provide a defence against the Scots. As Exeter puts it, 'government … | Put into parts, doth keep in one consent, | Congreeing in a full and natural close, | Like music' (1.2.180–183). The music analogy leads some editors to spell 'consent' as 'concent' in light of the *OED* definition of the latter as 'Harmony of sounds; accord or concord of several voices or parts; playing or singing together in harmony'; indeed, the *OED* itself gives this speech as an instance of the word's figurative application (*OED* 'concent', *n.*). However, in light of the imagery Canterbury deploys in his response, it may be legitimate to see the word as gesturing also towards the term 'concentric', that is, 'Of geometric figures or objects: having a common centre or central axis':

> Therefore doth heaven divide
> The state of man in divers functions,
> Setting endeavour in continual motion;
> To which is fixèd, as an aim or butt,
> Obedience. (1.2.183–187)

Canterbury insists that the various estates within a society share the common goal of obedience, like archers aiming for the same target though they stand in different positions. This is an image to which, after his Virgilian account of the honey bees, he will return:

> I this infer:
> That many things, having full reference
> To one consent, may work contrariously.
> As many arrows, loosèd several ways,
> Fly to one mark, as many ways meet in one town,
> As many fresh streams meet in one salt sea,
> As many lines close in the dial's centre,
> So may a thousand actions once afoot
> End in one purpose ... (1.2.204–212)

The monarch's *consent* is also the *centre* at which different endeavours *converge*, like arrows aiming for a target or the markings on the face of a clock or sundial. The way Canterbury deploys these circular images of dials and archery butts, which stresses the coming together of lines at a central point, reflects the classic definition of a circle by the Greek geometrician Euclid: in Henry Billingsley's 1570 translation, 'A circle is a plain figure, contained under one line, which is called a circumference, unto which all lines drawn from one point within the figure and falling upon the circumference thereof are equal the one to the other.'[39] A circle is understood not simply as a line that travels at a regular distance around a point, but as figure in which a potentially infinite number of equal lines converge at that point—like the 'many arrows' or 'many lines' organized around the will of the sovereign. Shakespeare could have found this definition in Billingsley's translation; or he could have found it in Recorde's *The Pathway to Knowledge* (1551), 'the first textbook of geometry for English readers', which rendered concepts and definitions derived from Euclid in English for the first time.[40] Recorde's text begins with a preface that stresses the practical uses of geometry, not just in contexts one might expect, such as building, navigation, painting, and clockmaking, but in the liberal arts, philosophy, medicine, jurisprudence, and government:

And although Plato do use the help of Geometry in all the most weighty matter of a commonwealth, yet it is so general in use, that no small things almost can be well done without it…. Lycurgus that chief lawmaker amongst the Lacedemonians, is most praised for that he did change the state of their commonwealth from the proportion Arithmetical to a proportion geometrical….[41]

Recorde's apology provides a rationale for Canterbury's use of geometrical imagery to describe the state. Not only does he envisage it as a circle (or series of concentric circles) whose radii converge in the monarch; his exhortation that Henry 'Divide your happy England into four, | Wherof take you one quarter into France' (1.2.214–215) imagines the nation as something that might be quartered as easily as a geometrical figure, rendering a complex spatial and political reality as a readily apprehended visual icon.

Circular and spherical images persist throughout *Henry V*, from the tennis balls with which the Dauphin mocks Henry to the 'gunstones' (1.2.282) Henry promises to send back in return, and from the English ordinance with their 'fatal mouths gaping on girded Harfleur' (3.0.27) to Henry's exhortation that his soldiers let their eyes 'pry through the portage of the head | Like the brass cannon' (3.1.10–11). This sequence culminates in the final scene in a bawdy exchange between Henry and the Duke of Burgundy following the King's wooing of Catherine, after which he complains that 'I cannot so conjure up the spirit of love in her that he will appear in his true likeness.' Burgundy responds:

If you would conjure in her, you must make a circle; if conjure up love in her in his true likeness, he must appear naked and blind. Can you blame her then, being a maid yet rosed over with the virgin crimson of modesty, if she deny the appearance of a naked blind boy in her naked seeing self? (5.2.286–288, 290–295)

Burgundy's reference to conjuring by making a circle alludes to the magic circle of the magician, an image implied earlier on in the Prologue's apology for the 'flat unraisèd spirits' appearing within 'this wooden O'. However, as Patricia Parker notes, the circle imagery also implies the anatomical 'female "O"', and consequently the act of sexual penetration by which Henry is expected to make a 'naked blind boy' (both the love-god Cupid and Henry's embryonic heir) appear within the (presumably) virginal Catherine through the phallic opening-up of a circle where there was none before.[42] It is tempting to wonder whether Burgundy's obscenely incongruous application of geometry in this context might have been enabled by the fact that, whereas Billingsley's translation of Euclid uses the word 'point' to describe *'an unity which hath position'* but 'neither hath length, breadth, nor thickness', such as the centre of a circle or the end of a line, Recorde's *Pathway* invariably uses the term 'prick', thereby leaving itself continually susceptible to double entendre.[43] According to Recorde's definition, 'when a prick standeth in the middle of a circle (as no circle can be made by compass without it) then is it called a *centre*'; the idea that the making of a circle requires a prick to stand at its centre seems to underlie the remark Burgundy is making.[44]

The misogyny of Burgundy's conceit consists not just in its depiction of the female body as an arena for the realization of phallic masculinity, however. In so far as it exploits the sexual suggestiveness of the circle, it also evokes the analogous figure of the zero, thanks to the way that (as Michele Jaffe puts it) the growing use of the zero 'retroactively changes the way that all round things are viewed'.[45] This effect is compounded by Shakespeare's repeated allusion to the use of the word 'nothing' as a slang term for the female genitals, as when the speaker of Sonnet 20 complains that nature has added to his androgynous addressee, first designed as a woman, 'one thing to my purpose nothing', or when Hamlet quibbles over Ophelia's response to his request to 'lie in your lap', 'I think nothing, my lord', 'That's a fair thought to lie between maids' legs' (3.2.107, 112–113). By this logic, Catherine's body is simultaneously diminished in patriarchal terms as bearing a zero, a nothing, no phallic 'thing', and aggrandized as the means by which royal masculinity is able to reproduce itself, just as the 'wooden O' referred to in the prologue is both an 'unworthy scaffold' and something with the power to magnify an actor into a king or multiply a soldier into an army. When Henry, lamenting the way his impending marriage is set to limit his imperial conquests, says that he 'cannot see many a fair French city for one fair French maid that stands in my way', the King of France insists that 'you see them perspectively, the cities turned into a maid—for they are all girdled with maiden walls that war hath never entered' (5.2.314–319). In so doing, he not only makes the commonplace equation between a woman and a walled town that lurked beneath Henry's threats against the 'pure maidens' of Harfleur (3.3.103); he imagines that Henry's view is mediated by some kind of optical device or 'perspective' that transforms many cities into a single female image (see *OED* 'perspective', *n.*, 2), reversing the act of multiplication carried out by the theatre in the prologue. In returning to the conceit of its opening speech, the play comes full circle, just as the epilogue's allusion to events from the reign of Henry's son, 'Which oft our stage hath shown' (Epilogue, 13), closes the circle of Shakespeare's historical sequence by looking forward to the *Henry VI* plays with which it began.

Noughts and Division in *King Lear*

Notwithstanding the downbeat ending of *Henry V*, which looks ahead to the King's early death and the territorial losses and civil strife that will bedevil the reign of his son, and notwithstanding the play's depiction of the horror of war and the involuntary dynastic marriage that concludes it, Shakespeare's use of mathematical imagery within this drama stresses the zero's power to multiply and treats the circle as a figure for wholeness and shared endeavour. When Shakespeare redeploys similar imagery in *King Lear*, however, it is used to diametrically opposite effect. Here, circles tend to be associated with pain and loss: the 'bleeding rings' of Gloucester's eyes (5.3.186), the 'wheel of fire' on which Lear says he is bound (4.6.45), the 'wheel … come full circled' that brings Edmund to his death (5.3.170; references are to the Quarto *History of King Lear* unless indicated otherwise). And in this play centred not around victory and expansion but crisis and loss, emphasis falls not on the zero's power to magnify but its power to negate and belittle, and not on multiplication, but division.

The concept of nothing is thematically important to *King Lear*: it is Cordelia's answer, 'Nothing, my lord', to Lear's request that she outdo her sisters' professions of love that sets in train her banishment and accelerates Lear's decline (1.1.81). The word chimes repeatedly through the play, as in Edmund's manipulation of his father with the statement that the letter he pretends to conceal is 'Nothing, my lord' (1.2.31) or Edgar's words when he resolves to disguise himself, 'Edgar I nothing am' (2.2.187). Edgar's paradox is indicative of the way, in all these instances, an apparent or supposed nothing proves to be very substantial: as Gloucester puts it to Edmund, 'The quality of nothing hath not such need to hide itself … if it be nothing I shall not need spectacles' (1.2.33–35). Shakespeare toys with the idea that a thing or person may be nothing and something at the same time, and the zero—a symbol that reifies nothing, turns it into something—provides a graphic way of illustrating this, most obviously in the Fool's lines to Lear, 'Now thou art an O without a figure. I am better than thou art, now. I am a fool; thou art nothing' (1.4.186–188). According to the Fool's mathematical analogy, just as a zero is made a number by the figure that stands to its left, so Lear was made what he was by his land and crown, and in giving them away, he has lost the attributes that defined him. This exchange builds on the Fool's offer earlier in the same scene:

> FOOL   Give me an egg, nuncle, and I'll give thee two
>   crowns.
> LEAR   What two crowns shall they be?
> FOOL   Why, after I have cut the egg in the middle and
>   eat up the meat, the two crowns of the egg. When
>   thou clovest thy crown i'th' middle and gavest away
>   both parts, thou borest thy ass o'th' back o'er the dirt.
>   Thou hadst little wit in thy bald crown when thou
>   gavest thy golden one away. (1.4.150–158)

These lines once again illustrate Jaffe's point about the introduction of the zero causing other round things to be viewed differently: the egg, the royal crown, and Lear's skull pass through this speech as images both of fullness and of emptiness, the crown variously suggesting both the symbol of authority for which the play's characters struggle and a zero, a nothing, the eggshell that remains once the contents have been eaten—a 'hollow crown', as King Richard II puts it (*Richard II*, 3.2.156). It is as if Lear's act of dividing the crown at the beginning of the play emptied it of value, and of the power to confer value on its possessor. Fittingly, the recipients of the two halves, Albany and Cornwall, receive with them two sisters who will both come to be associated with zeroes: Goneril, whom Lear describes as 'naught' (2.2.296), and Regan, whom the tormented Gloucester addresses as 'Naughty lady' (3.7.36). In both instances, 'naught' serves as a complex pun, simultaneously criticizing the morally disreputable, 'naughty' behaviour of the sisters, devaluing them as 'nought', and alluding to the figure of the zero that symbolizes their ungoverned sexuality.

Lear's invitation to Albany and Cornwall, 'This crownet part betwixt you' (1.1.131), emblematizes a second mathematical theme that runs through the play, namely that of division. Not only is the play premised on the notion of a king dividing his realm; it also depicts a world in which 'Love cools, friendship falls off, brothers divide' (1.2.106–107) and later on there is more specific 'division | ... 'twixt Albany and Cornwall' (3.1.18–20). However, division is an ambivalent term, as the obtuse Scholar in Recorde's *Ground of Arts* unwittingly demonstrates:

> SCHOLAR   Well sir, then in Division I pray you to
> instruct me. But me thinketh by the name of it that
> it should be all one with Multiplication: for I call
> that Division, when anything is parted into divers
> and many parts.
> MASTER   You take it, as it is taken commonly, howbeit
> if you mark well you shall perceive that it is quite
> contrary to Multiplication, and doth not part one
> thing or few things into many, but contrary ways it
> bringeth many parcels into few, but yet so, that these
> few taken together, are equal in value to the other
> many.[46]

The Scholar believes that division is the same thing as multiplication, because it involves turning one thing into many things. He does seem to have a point: if you divide one thing you end up with two things, which appears to multiply the number of things you started off with. (In this respect he anticipates the Chorus in *Henry V*, whose exhortation, 'Into a thousand parts divide one man' actually envisions a multiplication: the audience are asked to make one man a thousand in their imaginations.) But in reality, dividing a thing produces not things, but only parts of things, as becomes apparent when the Master invites the Scholar to divide 365 by 28. The Master shows how this gives 13 and one remaining, bringing the sum to an end—'except', as the Scholar points out, 'you would part the 1 that remaineth into 28 parts'. 'I will let that pass now', replies the Master, 'and will hereafter teach you peculiarly of broken number called fractions.'[47]

The distinction that the Master sets out between dividing numbers and working with 'broken numbers' is relevant to *King Lear* because of the play's textual ambiguity over what Lear actually envisages at the outset. Is it, as per the Quarto, a 'division of the kingdoms' (*History*, 1.1.4), a separation of things that already have distinct identities? Or is it, as per the Folio, a 'division of the kingdom' (*Tragedy*, 1.1.4) into three fractions each of which is merely part of a whole? This ambiguity as to what is or are being divided is accompanied by a degree of contradiction over the equality of the division. On the one hand Lear claims to be giving Regan an 'ample third of our fair kingdom, | No less in space, validity, and pleasure | Than that confirmed on Goneril' (1.1.74–76). This seems to imply a division into three equal parts. But when he offers Cordelia 'a third more opulent than your sisters', mathematics appear to have fallen by the wayside: Goneril's and Regan's thirds add up to a two-thirds that somehow leaves a bit more than a third available for Cordelia.

Of course, the real problem here is the collision of mathematical precision with geopolitical reality. Back in *Henry V*, Canterbury advised the King to 'Divide your happy England into four' as if that were a straightforward task; but when it comes to splitting up a land mass, factors such as irregular shape, topographical variation, and uneven distribution of resources make equal division impossible. Some things are resistant to quantification: not only national entities but, crucially to the play's opening scene, filial love. That does not, however, prevent Lear from demanding that his daughters 'Tell me … | Which of you shall we say doth love us most' (1.1.45–46). Nor does it stop Regan from responding to Goneril's claim to love her father 'Dearer than eyesight, space, or liberty; | Beyond what can be valued, rich or rare; | No less than life; with grace, health, beauty, honour' by asserting that 'she names my very deed of love— | Only she came short' (1.1.51–53, 66–67). It is as if any abstract quality can retroactively be turned into a quantity simply by adding more to it. For his part, Lear's later attempt to gauge his daughters' love on the basis of how many soldiers they will allow in his retinue takes this way of thinking to pathetic extremes: 'Thy fifty yet doth double five-and-twenty, | And thou art twice her love' (2.2.418–419).

The play sets up the impulse to quantify against the figure of Cordelia, who not only refuses to play this particular game, but is valued by France in a way that defies Lear's insistence that 'her price is fallen' (1.1.196). France contends that Cordelia's worth is not a number that can be raised or lowered through Lear's provision or denial of a dowry; 'She is herself a dowry' (1.1.241), 'unprized precious' (259), both unvalued and beyond price. Unlike the unkinged Lear, 'an O without a figure', or her 'naughty' sisters, Cordelia has a value that is both intrinsic and absolute, which makes her death at the end of the play—an apocalyptic 'image' of the 'promised end'—all the more harrowing (5.3.259–260). As Shankar Raman says of the deaths of Mamillius and Antigonus in *The Winter's Tale*, in an essay that finds in that play another Shakespearean exploration of the paradoxical attributes of the zero, 'There are deaths … that no accounting can ever be fully adequate to'.[48]

Conclusion: 'The Circumference of So Many Ears'

It would be tempting to read the insistence in *King Lear* that some things have a value that cannot be reduced to number as a broader attack on quantification in general, as represented by (for example) the accounting practices with which Shakespeare would have been familiar, or (more tenuously) the idea of the universe as apprehensible in mathematical terms that some early modern scientists came to regard as a central tenet. That would, however, be misleading. As Paula Blank points out, in the wake of her sisters' profession of limitless love for their father Cordelia portrays her own love for him as a measurable portion of affection: 'Haply when I shall wed | That lord whose hand must take my plight shall carry | Half my love with him' (1.1.92–94), the other half presumably staying with Lear. Any division of the play's characters into sympathetic ones who resist quantification and less sympathetic ones who resist it is therefore problematic. Furthermore, in the final scene Albany seems to offer the prospect of measured rewards and punishments when he restores Edgar and Kent to

> your rights,
> With boot and such addition as your honours
> Have more than merited. All friends shall taste
> The wages of their virtue, and all foes
> The cup of their deservings. (5.3.297–299)

The implication is that justice, measure for measure, requires some kind of proportionality, however hard that may be to identify. Blank rightly notes that '*King Lear* does not unequivocally refute arithmetic and the effort to evaluate human worth or to credit it through material rewards': the play combines scepticism about quantification with an acknowledgement of its necessity.[49]

A similar sense of irresolution surrounds the play's treatment of the questions of Providence and planetary influence that were discussed in Chapter 1. One example is the way Edmund's reasonable-sounding insistence that the 'late eclipses in the sun and moon' noted by his father (1.2.103) have nothing to do with current political and family disturbances—'as if we were villains on necessity, fools by heavenly compulsion, knaves, thieves, and treacherers by spherical pre-dominance' (116–118)—sits alongside other characters' propensity to see a divinely shaped order in events. 'The dark and vicious place where thee he got | Cost him his eyes', Edgar tells him of Gloucester (5.3.168–169), interpreting their father's blinding as a harsh and belated punishment for his adultery. Remarks like those of Edgar, which seek to find a logic in pain and suffering, might seem simplistic given the play's bleak outcome; but *King Lear* does not straightforwardly reject the possibility that human events might be ordered by supernatural means. Coincidentally or not, the eclipses cited by Gloucester do, in fact, precede family collapse and societal breakdown, while Edmund's sarcastic observation, 'My father compounded with my mother under the Dragon's Tail and my nativity was under Ursa Major, so that it follows I am rough and lecherous' (1.2.123–126) is not entirely incompatible with his characterization in the play. *King*

Lear does not force audiences to choose between a Lucretian understanding of the universe and one that allows for the gods, or the stars, shaping human actions and outcomes.

This openness, as the preceding chapters of this book have repeatedly indicated, is characteristic of the way Shakespeare treats scientific themes. Chapter 1 showed how the Histories anticipate *Lear* in that they depict characters who read astronomical and meteorological phenomena as divine warnings, and other characters who emphasize the natural causes that produce them, without resolving the difference between those perspectives. The response of the Welsh soldiers to the meteors in *Richard II*, for example, can variously be seen as superstition, as a legitimate response to a celestial sign that accurately portends Richard's downfall, and as a combination of the two—a superstitious response that, by leading to the denial of military support to Richard, brings his downfall into effect. Shakespeare's preference for leaving ambiguities like these unresolved might be attributed to the playwright's own beliefs—to the kind of epistemological scepticism that is evident in *Hamlet*. But a less personalized explanation might cite aesthetic considerations: leaving open the possibility of a supernatural agent directing historical events is arguably more intriguing and unsettling than presenting either a pre-determined or a merely secular universe. As this book has argued, the nature of the dramatic form itself encourages a dialogic treatment of scientific questions—in this instance, one that embraces ambiguity and multiple causalities. Moreover, the Histories' exploration of questions of Providence in relation to meteorological events gains another dimension from their being staged in venues whose use of special effects mimicked God's control of the weather.

The juxtaposition of multiple competing explanations is not, however, peculiar to Shakespeare's dramatic works. As Chapter 2 showed, the Sonnets are both preoccupied with themes of matter and substantiality, and informed by divergent models of what matter is and how it behaves in ways that reflect the influence of Ovid, of alchemical ideas, and of the theology of the Eucharist. Their derivation of scientific ideas from poetic, mystical, and religious sources exemplifies the extent to which the distinction between scientific and non-scientific discourses that seems obvious to the modern world was much less clear in Shakespeare's time, and the blurring of this boundary is particularly evident in the case of Lucretius, whose poetry influenced seventeenth-century scientists like Gassendi and Newton and may have reached Shakespeare via the writings of Montaigne. Distinctions between science and magic were no less hazy, as is indicated by the way in which the Neoplatonic magus Prospero in *The Tempest* seems to prefigure the Baconian scientist. The fundamentally theatrical nature of the power he exerts can be read as offering a sceptical critique of science, such that what looks like mastery of the elements is merely illusion. A more optimistic reading might identify the way Prospero's island serves as an analogue for the playhouse as a gesture towards the theatre's own capacity to be a place for the making of scientific knowledge, where speculations on the nature of matter were voiced by performers whose bodies could be mercurial and protean or obdurately hard to shift.

Another type of space with which the early modern theatre had affinities was discussed in Chapter 3, namely the celebrated anatomy theatres of Bologna, Leiden, and Padua (and to a lesser extent the College of Physicians and the Barber-Surgeons' Hall in London). However, while playhouses, too, occasionally staged scenes of bloody dismemberment, Shakespeare's plays are frequently preoccupied by what cannot be known simply by examining or penetrating the body, be that marital chastity, the paternity of a child, or the contents of the mind. Strikingly, one prominent instance where opening the body does provide a form of revelation comes not in a play, but in the narrative poem *Lucrece*, where Shakespeare exploits the possibilities of a non-dramatic form by graphically depicting a suicide whose bloody aftermath confirms Lucrece's story. The poem's reliance on narration also enables Shakespeare to depict in detail the way the passions affect the body: as in the plays, where emotional change is frequently conveyed through the language of the humours, the workings of the soul are very much an embodied phenomenon.

It was not only the bodies of fictional characters that were acted upon in early modern playhouses, though: dramas provoked emotional responses from playgoers that could be manifested in the body in the form of laughing and weeping, as well as placing them in physical proximity that exposed them to infection. As the current chapter has shown, *Henry V* enlists playgoers in a different kind of activity: a form of collective mathematics whereby the bodies of the actors are combined with the zero suggested by the playhouse's shape to multiply them into armies. Shakespeare capitalizes on the contradictory qualities of the zero, which can augment and can negate, in ways that are evident in *Henry V* and in *King Lear*; at the same time, both plays exploit the visual similarity of the nought and the circle, an image that is the basis for rhetorical conceits on subjects ranging from political theory to misogynistic satire.

Shakespeare was not alone in capitalizing on the roundness of the playhouse when describing the relationship between actors and their audiences, and the theatre's ability to offer nature up for observation. Another writer, possibly Shakespeare's fellow playwright and Globe dramatist John Webster, wrote as follows in his description of 'An excellent Actor' added to the 1615 edition of Thomas Overbury's *The Wife*:

Whatsoever is commendable in the grave Orator, is most exquisitely perfect in him; for by a full and significant action of body, he charms our attention: sit in a full Theatre, and you will think you see so many lines drawn from the circumference of so many ears, whiles the *Actor* is the *Centre*. He doth not strive to make nature monstrous, she is often seen in the same Scene with him, but neither on Stilts nor Crutches....[50]

This passage makes a claim for the theatre's ability to depict nature accurately and without exaggeration or diminution, 'neither on Stilts nor Crutches'. At the same time, the author exploits, like Shakespeare, the shape of the outdoor playhouse to position the actor at the centre of a circle whose circumference is made up of admiring auditors. This conceit is recalled in the later observation that 'a man of a deep thought might apprehend, the Ghosts of our ancient *Heroes* walked again': the idea that the theatre might be a space in which the dead are revived implies the image of the conjuror's magic circle that has already been noted in Shakespeare's reference in *Henry V* to 'flat unraisèd spirits'. Even more pertinently to this chapter, however, the description of 'so many lines drawn from the circumference of so many ears, whiles the *Actor* is the *Centre*' alludes to the Euclidean definition of a circle recycled, in very different ways, by Canterbury and Burgundy: 'a plain figure, contained under one line, which is called a circumference, unto which all lines drawn from one point within the figure and falling upon the circumference thereof are equal the one to the other'. For the author of 'An excellent Actor' the theatre was, amongst other things, a geometrical phenomenon, in a way that physically involved audiences in the discussions of value, measurement, and social structure that plays like *Henry V* and *King Lear* used mathematics to illustrate.

Although the writer does not specify any playhouse in particular as the setting of the scene he describes, it would make sense for it to be the Globe, a theatre whose name claimed for the polygonal building a symbolically charged circularity it did not in fact possess. That name linked the theatre both with the earth and the spherical cosmos, and also with the terrestrial and celestial globes that were used to map and to observe them. In so doing it made the playhouse both a universe in miniature, and a technology that might be used to lay the universe open to view. This is a bold claim: as a means of representing nature, the theatre is a very different kind of instrument from a finely calibrated sphere or polished telescope. However, as this book has argued, the playhouse, and theatrical performance itself, were crucial factors in Shakespeare's exploration of scientific themes. Not only did the dialogue and conflict integral to drama facilitate provisionality, experiment, and play; the stubborn physicality of actors, the contrasting potential for illusion and change, the use of pyrotechnics and other effects, and the bringing together of audiences in a collective showing-space all added a dimension to his work that would have been absent from purely literary texts. For all its reliance on fictionality and pretence—qualities that might seem to be at odds with scientific values and aspirations—the Shakespearean theatre deserves to be seen as a significant venue in which early modern science was conducted.

INTRODUCTION

1. On *Doctor Faustus* as the 'ur-science play', see Kirsten Shepherd-Barr, *Science on Stage: From 'Doctor Faustus' to 'Copenhagen'* (Princeton: Princeton University Press, 2006), p. 17.

2. John Donne, 'An Anatomy of the World: The First Anniversary', in *The Complete English Poems*, ed. A. J. Smith (London: Penguin, 1971, repr. 1986), p. 276.

3. C. P. Snow, 'The Rede Lecture, 1959', in C. P. Snow, *The Two Cultures* (Cambridge: Cambridge University Press, 1993 (1965)), pp. 1–51, 14–15.

4. Sally Weale, 'Funding cuts to go ahead for university arts courses in England despite opposition', *Guardian*, 20 July 2021; 'More young people are taking STEM subjects than ever before', Education Hub, gov.uk, 2 February 2021.

5. For an overview, see Clarisse Loughrey, 'William Shakespeare: How the UK is celebrating his 400th anniversary', independent.co.uk, 22 April 2016.

6. Aristotle, *Nicomachean Ethics*, trans. H. Rackham (Cambridge, Mass: Harvard University Press, 1926), book 6, chapters iii–iv. See Elizabeth Spiller, 'Shakespeare and the Making of Early Modern Science: Resituating Prospero's Art', *South Central Review* 26 (2009): 24–41, 27.

7. John Dee, Preface to Euclid, *The Elements of Geometrie*, trans. Henry Billingsley (London, 1570), sig. b1^{r}.

8. St Thomas Aquinas, *Summa Theologiae*, vol. 1: *Christian Theology*, trans. Thomas Gilby (London: Eyre and Spottiswoode; New York: McGraw-Hill, 1964), Part 1, Question 1, Article 5 (p. 19).

9. Katharine Park and Lorraine Daston, 'Introduction: The Age of the New', in *The Cambridge History of Science*, vol. 3: *Early Modern Science*, ed. Katharine Park and Lorraine Daston (Cambridge: Cambridge University Press, 2006), pp. 1–17, p. 4.

10. See Ann Blair, 'Natural Philosophy', in Park and Daston eds, pp. 365–406, p. 366.

11. Michael H. Shank and David C. Lindberg, 'Introduction', in *The Cambridge History of Science*, vol. 2: *Medieval Science*, ed. Michael H. Shank and David C. Lindberg (Cambridge: Cambridge University Press, 2013), pp. 1–26, p. 7.

12. See Seb Falk, *The Light Ages: A Medieval Journey of Discovery* (London: Allen Lane, 2020), p. 8.

13. Aristotle, *On Coming-to-be and Passing Away*, book 2, chapter 3, in *On Sophistical Refutations. On Coming-to-be and Passing Away. On the Cosmos*, trans. E. S. Forster and D. J. Furley (Cambridge, Mass: Harvard University Press, 1955).

14. Francis Bacon, *The Novum Organum . . . Epitomiz'd*, trans. M. D. (London, 1576), pp. 2, 4.

15. Mary Thomas Crane, *Losing Touch With Nature: Literature and the New Science in Sixteenth-Century England* (Baltimore: Johns Hopkins University Press, 2014), p. 2.

16. Herbert Butterfield, *The Origins of Modern Science*, rev. ed. (London: G. Bell and Sons, 1957), pp. 7–8.

17. Alexandre Koyré, 'Galileo and the Scientific Revolution of the Seventeenth Century', *Philosophical Review* 52 (1943): 333–348; A. Rupert Hall, *The Scientific Revolution, 1500–1800: The Formation of the Modern Scientific Attitude* (London: Longmans, Green, 1954); J. D. Bernal, *Science in History*, vol. 2: *The Scientific and Industrial Revolutions* (London: Watts, 1954). See Steven Shapin, *The Scientific Revolution* (Chicago: University of Chicago Press, 1996), p. 2.

18. Robert Recorde, *The Castle of Knowledge* (London, 1556), sig. F2^v.

19. Adam Max Cohen, *Shakespeare and Technology: Dramatizing Early Modern Technological Revolutions* (New York: Palgrave Macmillan, 2006), p. 4; Deborah E. Harkness, *The Jewel House: Elizabethan London and the Scientific Revolution* (New Haven: Yale University Press, 2007).

20. Shapin, pp. 3–4.

21. Thomas Sprat, *The History of the Royal-Society of London, for the Improving of Natural Knowledge* (London, 1667), pp. 19, 22, 1.

22. See Peter Dear, *Revolutionizing the Sciences: European Knowledge and its Ambitions, 1500–1700* (Houndmills: Palgrave, 2001), pp. 34–35, 38.

23. See David C. Lindberg and Robert S. Westman, 'Introduction', in *Reappraisals of the Scientific Revolution*, ed. David C. Lindberg and Robert S. Westman (Cambridge: Cambridge University Press, 1990), p. xvii–xxvii, p. xix, and Gary Hatfield, 'Metaphysics and the New Science', in Lindberg and Westman eds, pp. 93–166, p. 108. The Platonic solids are the tetrahedron, cube, octahedron, dodecahedron, and icosahedron.

24. Edward Grant, *Much Ado about Nothing: Theories of Space and Vacuum from the Middle Ages to the Scientific Revolution* (Cambridge: Cambridge University Press, 1981), p. xi.

25. Falk, p. 55.

26. Sprat, p. 45.

27. Shapin, p. 8.
28. Mary Lindemann, *Medicine and Society in Early Modern Europe*, 2nd edn (Cambridge: Cambridge University Press, 2010), p. 3.
29. David J. Hess, *Science Studies: An Advanced Introduction* (New York: New York University Press, 1997), p. 1.
30. Dear, p. 2.
31. Ofer Gal, *The Origins of Modern Science: From Antiquity to the Scientific Revolution* (Cambridge: Cambridge University Press, 2021), p. 3.
32. Carla Mazzio, 'Shakespeare and Science, c. 1600', *South Central Review* 26 (2009): 1–23, 6.
33. Howard Marchitello and Evelyn Tribble, 'Introduction', in *The Palgrave Handbook of Early Modern Literature and Science*, ed. Howard Marchitello and Evelyn Tribble (London: Palgrave Macmillan, 2017), pp. xxv–xlvi, p. xxvi. Marchitello and Tribble cite works including Charles M. Coffin, *John Donne and the New Philosophy* (New York: Columbia University Press, 1937); William Empson, *Essays on Renaissance Literature*, vol. 1: *Donne and the New Philosophy*, ed. John Haffenden (Cambridge: Cambridge University Press, 1993); and Marjorie Hope Nicolson, *The Breaking of the Circle: Studies in the Effect of the 'New Science' upon Seventeenth-Century Poetry* (Evanston: Northwestern University Press, 1950).
34. Juliet Cummins and David Burchell, 'Introduction', in *Science, Literature and Rhetoric in Early Modern England* (Aldershot: Ashgate, 2007), pp. 1–12, p. 2.
35. *The Defence of Poesy*, in Sir Philip Sidney, *A Critical Edition of the Major Works*, ed. Katherine Duncan-Jones (Oxford: Oxford University Press, 1989), p. 213.
36. Sprat, p. 62.
37. Francis R. Johnson, *Astronomical Thought in Renaissance England: A Study of the English Scientific Writings from 1500 to 1645* (Baltimore: Johns Hopkins University Press, 1937; repr. New York: Octagon, 1968), p. 10.
38. See Nancy G. Siraisi, *Medieval and Early Renaissance Medicine: An Introduction to Knowledge and Practice* (Chicago: University of Chicago Press, 1990), pp. 67–68.
39. See Patrick Cheney, *Shakespeare, National Poet-Playwright* (Cambridge: Cambridge University Press, 2004), p. 17.
40. Katherine Walker, *Shakespeare and Science: A Dictionary* (London: Arden Shakespeare, 2022), p. 3.
41. Henry S. Turner, *The English Renaissance Stage: Geometry, Poetics, and the Practical Spatial Arts, 1580–1630* (Oxford: Oxford University Press, 2006), p. 29.

42. Jean Feerick, 'Matter, Nature, Cosmos: The Scientific Art of the Early Modern English Stage', in *The Arden Handbook of Shakespeare and Early Modern Drama*, ed. Michelle M. Dowd and Tom Rutter (London: Bloomsbury, 2023), pp. 201–217, p. 204.
43. Cohen, p. 51; Recorde, sig. C6^r.
44. The earliest instance of the term 'anatomy theatre' that I have been able to find is in the Latin title of a 1609 engraving of the anatomical theatre at Leiden, 'Theatrum anatomicum van Leidse Academie'. Catalogue reference FMH 1284-D, Rijksmuseum, Amsterdam.

CHAPTER 1

1. Christopher Marlowe, *Doctor Faustus* (A-text), in *The Complete Plays*, ed. Frank Romany and Robert Lindsey (London: Penguin, 2003), Chorus 2.1–7.
2. Barnabe Googe, *The Zodiake of Life Written by the Godly and Zealous Poet Marcellus Pallingenius Stellatus* (London, 1565), sigs PP7^{r-v}.
3. Seb Falk, *The Light Ages: A Medieval Journey of Discovery* (London: Allen Lane, 2020), pp. 75–76.
4. Falk, pp. 171, 242–243, 263–265.
5. Aristotle, *Meteorologica*, trans. H. D. P. Lee (Cambridge, Mass: Harvard University Press, 1952), book 1, chapter 9; book 1, chapter 4; book 2, chapter 4; book 2, chapter 8.
6. Aristotle, *The Physics*, trans. Philip H. Wicksteed and Francis M. Cornford, rev. ed. (Cambridge, Mass: Harvard University Press, 1957), book 3, chapter 5.
7. S. K. Heninger, *A Handbook of Renaissance Meteorology: With Particular Reference to Elizabethan and Jacobean Literature* (Durham, N. C.: Duke University Press, 1960), p. 204.
8. Ann Blair, 'Natural Philosophy', in *The Cambridge History of Science*, vol. 3: *Early Modern Science*, ed. Katharine Park and Lorraine Daston (Cambridge: Cambridge University Press, 2006), pp. 365–406, pp. 366–369.
9. Mary Thomas Crane, *Losing Touch with Nature: Literature and the New Science in Sixteenth-Century England* (Baltimore: Johns Hopkins University Press, 2014), pp. 2, 22, 75.
10. Kristen Poole, *Supernatural Environments in Shakespeare's England: Spaces of Demonism, Divinity and Drama* (Cambridge: Cambridge University Press, 2011), pp. 4, 10.
11. Steve Mentz, 'Hurricanes, Tempests and the Meteorological Globe', in *The Palgrave Handbook of Early Modern Literature and Science*, ed. Howard

Marchitello and Evelyn Tribble (London: Palgrave, 2017), pp. 257–276, p. 257.

12. Steve Mentz, 'Strange Weather in *King Lear*', *Shakespeare* 6 (2010): 139-52.

13. Gabriel Egan, *Green Shakespeare: From Ecopolitics to Ecocriticism* (London: Routledge, 2006), pp. 132–171; Todd A. Borlik, *Ecocriticism and Early Modern English Literature: Green Pastures* (New York: Routledge, 2011), pp. 128, 120–121.

14. Leslie Thomson, 'The Meaning of *Thunder and Lightning*: Stage Directions and Audience Expectations', *Early Theatre* 2 (1999): 11–24, 11.

15. Gwilym Jones, *Shakespeare's Storms* (Manchester: Manchester University Press, 2015), p. 16.

16. Sophie Chiari, *Shakespeare's Representation of Weather, Climate and Environment: The Early Modern 'Fated Sky'* (Edinburgh: Edinburgh University Press, 2019), pp. 57–79, pp. 66–67, citing William Vaughan, *Approved Directions for Health* (London, 1612), p. 69.

17. Poole, pp. 136–167.

18. Richard Hooker, *Of the Lavves of Ecclesiasticall Politie* (London, 1593), p. 50

19. John Calvin, *The Institution of Christian Religion*, trans. Thomas Norton (London, 1561), chapters 16.3, 16.9.

20. These range from E. M. W. Tillyard, who finds across the two tetralogies a long-term working out of divine justice in which the deposition of Richard II is punished by decades of civil strife that are ultimately resolved by the ascendancy of the Tudors, to Phyllis Rackin, who discerns a more self-conscious scrutiny of the whole concept. See E. M. W. Tillyard, *Shakespeare's History Plays* (London: Chatto and Windus, 1944), pp. 320–321; Phyllis Rackin, *Stages of History: Shakespeare's English Chronicles* (London: Routledge, 1991), pp. 5–12.

21. Calvin, chapter 16.5.

22. Arthur Golding, *A Discourse vpon the Earthquake that Hapned through this Realme of Englande, and Other Places of Christendom, the Sixt of Aprill. 1580 betwene the Houres of Fiue and Six in the Euening* (London, 1580), sigs B1^r-B1^v.

23. Thomas Churchyard, *A Warning for the Wise, a Feare to the Fond, a Bridle to the Lewde, and a Glasse to the Good* (London, 1580), sigs A2^v-A3^r.

24. Thomas Twyne, *A Shorte and Pithie Discourse, Concerning the Engendring, Tokens, and Effects of All Earthquakes in Generall* (London, 1580), sigs A1^{r-v}, A3^v.

25. William Fulke, *A Goodly Gallerye … to Behold the Naturall Causes of All Kynde of Meteors* (London, 1563), sigs A4^v, F4^r.

26. Gabriel Harvey and Edmund Spenser, *Three Proper, and Wittie, Familiar Letters: Lately Passed betvveene Tvvo Vniuersitie Men: Touching the Earthquake in Aprill Last, and Our English Refourmed Versifying* (London, 1580), pp. 16–20.
27. Edward Hall, *The Vnion of the Two Noble and Illustrate Famelies of Lancastre and Yorke* (London, 1548), sig. JJ6^v.
28. *Tamburlaine the Great, Part One*, in *The Complete Plays*, ed. Romany and Lindsey, 1.2.185.
29. George Peele, *The Troublesome Raigne of Iohn King of England* (London, 1591), Part 2, sigs A2^r, E2^v; Part 1, sig. G2^v.
30. Taylor and Loughnane also suggest that Parts 2 and 3 are collaborations between Shakespeare, Marlowe and others. See Gary Taylor and Rory Loughnane, 'The Canon and Chronology of Shakespeare's Works', in *The New Oxford Shakespeare: Authorship Companion*, ed. Gary Taylor and Gabriel Egan (Oxford: Oxford University Press, 2017), pp. 418–603, pp. 494–514.
31. Marlowe, *Tamburlaine*, Part 2, 5.3.249.
32. R. A. Foakes and R. T. Rickert eds, *Henslowe's Diary*, 2nd edn (Cambridge: Cambridge University Press, 2002), pp. 16–19.
33. Thomas Lodge and Robert Greene, *A Looking Glasse for London and England* (London, 1594), sigs C2^v, G2^r.
34. Hall, sig. Gg4^r.
35. Heninger, pp. 135–136.
36. Fulke, sigs F1^v-F2^v.
37. George Peele, *The Troublesome Raigne of Iohn King of England* (London, 1591), Part 2, sig. B3^r.
38. Raphael Holinshed and others, *The First and Second Volumes of Chronicles* (London, 1587), vol. 3 (*sic*), pp. 163, 180.
39. See Steven Shapin, *A Social History of Truth: Civility and Science in Seventeenth-Century England* (Chicago: University of Chicago Press, 1994), throughout, but for example (on the unreliability of non-gentlemen) pp. 93–94.
40. Fulke, sig. B8^r.
41. Aristotle, *Meteorologica*, book 2, chapter 8.
42. Marlowe, *Tamburlaine*, Part 1, 4.2.41–46.
43. Aristotle, *Meteorologica*, book 2, chapter 9.
44. Tom MacFaul, *Shakespeare and the Natural World* (Cambridge: Cambridge University Press, 2015), p. 2; Poole, p. 2.
45. Howard Marchitello, *The Machine in the Text: Science and Literature in the Age of Shakespeare and Galileo* (Oxford: Oxford University Press, 2011), pp. 57, 59. Natalie Elliot also offers an accessible account of *Hamlet* and

Copernicanism in 'Shakespeare's Worlds of Science', *The New Atlantis* 54 (Winter 2018): 31–50.

46. Carla Mazzio, 'The History of Air: *Hamlet* and the Trouble with Instruments', *South Central Review* 26 (2009): 153–196, 161.

47. Mazzio, 171.

48. Poole, p. 21.

49. Crane, p. 149.

50. John Gascoigne, 'A Reappraisal of the Role of the Universities in the Scientific Revolution', in *Reappraisals of the Scientific Revolution*, ed. David C. Lindberg and Robert S. Westman (Cambridge: Cambridge University Press, 1990), pp. 207–260, p. 230.

51. Marchitello, text for note 28 (p. 66), unpaginated in ebook.

52. Robert Recorde, *The Castle of Knowledge* (London, 1556), sig. O5^r.

53. Francis R. Johnson, *Astronomical Thought in Renaissance England: A Study of the English Scientific Writings from 1500 to 1645* (Baltimore: Johns Hopkins Press, 1937), p. 132.

54. Thomas Digges, *A Perfit Description of the Caelestiall Orbes*, appended to Leonard Digges, *A Prognostication Euerlastinge* (London, 1576), sigs. N2^v, N4^r.

55. Michel de Montaigne, *The Essayes or Morall, Politike and Millitarie Discourses*, trans. John Florio (London, 1603), p. 331.

56. Recorde, sig. A2^v.

57. John Donne, 'An Anatomy of the World: The First Anniversary', in *The Complete English Poems*, ed. A. J. Smith (London: Penguin, 1971), ll. 207–208.

CHAPTER 2

1. Aristotle, *On Coming-to-be and Passing Away*, book 2, chapters 2–3, in *On Sophistical Refutations. On Coming-to-be and Passing Away. On the Cosmos*, trans. E. S. Forster and D. J. Furley (Cambridge, Mass: Harvard University Press, 1955).

2. Ovid, *Metamorphoses*, trans. Arthur Golding, ed. Madeleine Forey (London: Penguin, 2002), 15.263–267.

3. Ovid, 15.183, 288–289, 201–203; see also Jonathan Bate, *Shakespeare and Ovid* (Oxford: Clarendon Press, 1993), p. 91.

4. Ovid, 15.76–80.

5. Ovid, 15.83; Colin Burrow, *Shakespeare and Classical Antiquity* (Oxford: Oxford University Press, 2013), p. 98.

6. Ovid, 15.202–203.

7. Ovid, 15.258, 279.

8. William R. Newman, 'From Alchemy to "Chymistry"', in *The Cambridge History of Science*, vol. 3: *Early Modern Science*, ed. Katharine Park and Lorraine Daston (Cambridge: Cambridge University Press, 2006), pp. 497–517, pp. 514–516. See also Charles Webster, *From Paracelsus to Newton: Magic and the Making of Modern Science* (Cambridge: Cambridge University Press, 1982).

9. Deborah E. Harkness, *The Jewel House: Elizabethan London and the Scientific Revolution* (New Haven: Yale University Press, 2007), p. 170.

10. Margaret Healy, *Shakespeare, Alchemy and the Creative Imagination: The Sonnets and 'A Lover's Complaint'* (Cambridge: Cambridge University Press, 2011), pp. 24, 67, 63.

11. Dympna Callaghan, 'Confounded by Winter: Speeding Time in Shakespeare's Sonnets', in *A Companion to Shakespeare's Sonnets*, ed. Michael Schoenfeldt (Malden: Blackwell, 2007), pp. 104–118, p. 111.

12. Newman, p. 504.

13. Harkness, pp. 61, 57–96.

14. For discussions of this topic in studies that address transubstantiation with reference to early modern poetry, see Sophie Read, *Eucharist and the Poetic Imagination in Early Modern England* (Cambridge: Cambridge University Press, 2013), esp. p. 15; Kimberly Johnson, *Made Flesh: Sacrament and Poetics in Post-Reformation England* (Philadelphia: University of Pennsylvania Press, 2014), esp. p. 13; Katherine Eggert, *Disknowledge: Literature, Alchemy, and the End of Humanism in Renaissance England* (Philadelphia: University of Pennsylvania Press, 2015), esp. p. 60.

15. St Thomas Aquinas, *Summa Theologiae*, vol. 58: *The Eucharistic Presence*, trans. William Barden (London: Eyre and Spottiswoode; New York: McGraw-Hill, 1965), Part 3, Question 75, Article 5 (pp. 73–77); Question 76, Article 1 (p. 97).

16. John Foxe, *The Pope Confuted*, trans. James Bell (London: Thomas Dawson for Richard Sergier, 1580), fols 100^r, 79^v. Conversely, the Catholic polemicist Thomas Harding complained that Protestant deniers of the Real Presence made Christ 'a minister of shadows, signs, tokens, and figures', and admitted 'figures for truth, tropes for the letter, shadows for things': *A Confutation of a Book Intituled An Apologie of the Church of England* (Antwerp: Jhon Laet, 1565), fols 7^v, 92^r.

17. On *Jack Juggler*, see William N. West, 'What's the Matter with Shakespeare?: Physics, Identity, Playing', *South Central Review* 26 (2009): 103–126, and Beatrice Groves, '"One man at one time may be in two placys": *Jack Juggler*, Proverbial Wisdom, and Eucharistic Satire', *Medieval and Renaissance Drama in England* 27 (2014): 40–56. For a discussion of Eucharistic imagery elsewhere in the Sonnets, see Rhema Hokama,

'Love's Rites: Performing Prayer in Shakespeare's Sonnets', *Shakespeare Quarterly* 63 (2012): 199–223.

18. Stephen Greenblatt, *Hamlet in Purgatory* (Princeton: Princeton University Press, 2001), p. 241.

19. Carla Mazzio, 'The History of Air: *Hamlet* and the Trouble with Instruments', *South Central Review* 26 (2009): 153–196 (169).

20. On Aristotle's ongoing 'intellectual domination of Western Europe and European colonies', see Charles B. Schmitt, *Aristotle and the Renaissance* (Cambridge, MA: Harvard University Press, 1983), p. 2.

21. Robert Hugh Kargon, *Atomism in England from Hariot to Newton* (Oxford: Clarendon Press, 1966), pp. 1–4.

22. Daniel Garber, 'Physics and Foundations', in *Early Modern Science*, ed. Park and Daston, pp. 21–69, p. 48.

23. See Gerard Passannante, *The Lucretian Renaissance: Philology and the Afterlife of Tradition* (Chicago: University of Chicago Press, 2011), pp. 27, 36.

24. For a lively description of the manuscript's rediscovery, see Stephen Greenblatt, *The Swerve: How the Renaissance Began* (London: Bodley Head, 2011), although some of Greenblatt's generalizations about the Middle Ages need to be treated with caution.

25. Jonathan Gil Harris, 'Atomic Shakespeare', *Shakespeare Studies* 30 (2002): 47–51 (48); Passannante, *Lucretian Renaissance*, pp. 104–119.

26. Passannante, *Lucretian Renaissance*, pp. 172, 199.

27. Robert M. Schuler, 'Francis Bacon and Scientific Poetry', *Transactions of the American Philosophical Society* 82.2 (1992): 1–65 (34–41); Kargon, pp. 24, 26, 29.

28. Kargon, pp. 3–4.

29. Christoph Meinel, 'Early Seventeenth-Century Atomism: Theory, Epistemology, and the Insufficiency of Experiment', *Isis* 79 (1988): 68–103 (77); Lucretius, *On the Nature of the Universe*, trans. Ronald Melville (Oxford: Clarendon Press, 1997), 2.391–397. Further references in the text are to this edition, which is lineated according to the Latin original.

30. David Jardine, *Criminal Trials, Supplying Copious Illustrations of the Important Periods of English History during the Reigns of Queen Elizabeth and James I*, vol. 1 (London: Nattali, 1847), p. 451. See also Kargon, p. 28.

31. Christopher Marlowe, *Doctor Faustus* (A-text), in *The Complete Plays*, ed. Frank Romany and Robert Lindsey (London: Penguin, 2003), 5.128.

32. Francis Meres, *Palladis tamia: Wits Treasury* (London, 1598), fol. 286^{v}.

33. Passannante in *Lucretian Renaissance* variously gives figures of 149 and 147 (pp. 105, 130). 'An Apologie of Raymond Sebond' is on pp. 252–351

of Michel de Montaigne, *The Essayes or Morall, Politike and Millitarie Discourses*, trans. John Florio (London, 1603); the discussion of atomist creation theory is on pp. 315–316.

34. See Suparna Roychoudhury, *Phantasmatic Shakespeare: Imagination in the Age of Early Modern Science* (Ithaca: Cornell University Press, 2018), pp. 88–91, 101–109.

35. Jonathan Pollock, 'Shakespeare and the Atomist Heritage', in *The Circulation of Knowledge in Early Modern English Literature*, ed. Sophie Chiari (Farnham: Ashgate, 2015), pp. 47–56, p. 55.

36. Robert N. Watson, 'False Immortality in *Measure for Measure*: Comic Means, Tragic Ends', *Shakespeare Quarterly* 41 (1990): 411–432 (415).

37. See Kargon, p. 29.

38. Mary Thomas Crane, *Losing Touch with Nature: Literature and the New Science in Sixteenth-Century England* (Baltimore: Johns Hopkins University Press, 2014), pp. 132, 138–139, 143.

39. Philipp Melanchthon, *Orations on Philosophy and Education*, trans. Christine Salazar, ed. Sachiko Kusukawa (Cambridge: Cambridge University Press, 1999), p. 176; quoted in Gerard Passannante, *Catastrophizing: Materialism and the Making of Disaster* (Chicago: University of Chicago Press, 2010), p. 91. In rationalizing his betrayal of his father as an instance of the natural law that 'The younger rises when the old do fall' (3.3.25), Edmund appropriates the Lucretian view that 'the old order always by the new / Thrust out gives way' (3.964–965).

40. Sean Ferrier and Lisa Walters, '*Hamlet* and Lucretian Anxiety', *Shakespeare* 18 (2022): 176–196, 180; Passannante, *Lucretian Renaissance*, p. 105.

41. West, pp. 103, 116.

42. William Shakespeare, *Hamlet*, ed. Ann Thompson and Neil Taylor (London: Arden Shakespeare, 2006), pp. 560, 562.

43. I take the notion of intertheatricality from Jonathan Gil Harris, *Untimely Matter in the Time of Shakespeare* (Philadelphia: University of Pennsylvania Press, 2009), where it refers to 'the untimely performance within Shakespeare's present of its theatrical pasts … and futures' by means of material phenomena such as the bodies of actors, costumes, and props (p. 68). The possible afterlife of Yorick's skull in *The Revenger's Tragedy* would be another example: see Lucy Munro, '"They eat each other's arms": Stage Blood and Body Parts', in *Shakespeare's Theatres and the Effects of Performance*, ed. Farah Karim-Cooper and Tiffany Stern (London: Bloomsbury Arden Shakespeare, 2013), pp. 73–93, p. 87.

44. See Pollock, 'Shakespeare and the Atomist Heritage', pp. 52–53.

45. For a survey, see Barbara A. Mowat, 'Prospero, Agrippa, and Hocus Pocus', *English Literary Renaissance* 11 (1981): 281–303 (282).

46. See, for example, Brian P. Copenhaver, 'Natural Magic, Hermeticism, and Occultism in Early Modern Science', in *Reappraisals of the Scientific Revolution*, ed. David C. Lindberg and Robert S. Westman (Cambridge: Cambridge University Press, 1990), pp. 261–301, and 'Magic', in *Early Modern Science*, ed. Park and Daston, pp. 518–540.

47. Mark A. Waddell, *Magic, Science, and Religion in Early Modern Europe* (Cambridge: Cambridge University Press, 2021), pp. 40–42.

48. George Molland, 'Bacon, Roger (c. 1214–1292?)', *ODNB*, 23 September 2004. Mowat (289–292) links Prospero with Bacon and notes other scholars who have done likewise.

49. Robert Recorde, *The Pathvvay to Knowledg Containing the First Principles of Geometrie* (London, 1551), sig. †3^v.

50. Francis Bacon, *Of the Proficience and Aduancement of Learning, Diuine and Humane* (London, 1605), fol. 24^r.

51. Elizabeth Spiller, 'Shakespeare and the Making of Early Modern Science: Resituating Prospero's Art', *South Central Review* 26 (2009): 24–41, 35, 31–32; Mickaël Popelard, 'Unlimited Science: The Endless Transformation of Nature in Bacon and Shakespeare's *The Tempest*', in *Spectacular Science, Technology and Superstition in the Age of Shakespeare*, ed. Sophie Chiari and Mickaël Popelard (Edinburgh: Edinburgh University Press, 2017), pp. 171–942, p. 184.

52. Spiller, pp. 31–32.

53. Denise Albanese, *New Science, New World* (Durham, NC: Duke University Press, 1996), p. 69.

54. Thomas Hariot, *A Briefe and True Report of the New Found Land of Virginia* (London, 1588), sigs E1^v, E4^r.

55. My formulation here is indebted to the oft-cited dictum, 'Any sufficiently advanced technology is indistinguishable from magic.' Arthur C. Clarke, letter to *Science*, 3812 (19 Jan. 1968): 255.

56. Hariot, sig. F2^r.

57. See B. J. Sokol, *A Brave New World of Knowledge: Shakespeare's* The Tempest *and Early Modern Epistemology* (Madison: Fairleigh Dickinson University Press; London: Associated University Presses, 2003), pp. 103–111.

58. Jenny C. Mann and Debapriya Sarkar, 'Introduction: "Capturing Proteus"', *Philological Quarterly* 98 (2019): 1–22, 8–9. See for example *The Wisdom of the Ancients*, in *The Works of Francis Bacon*, 14 vols. ed. James Spedding, Leslie Ellis, and Douglas Denon Heath et al. (London: Longman, 1861–1879), 6: 725.

CHAPTER 3

1. Sir George Clark and A. M. Cooke, *A History of the Royal College of Physicians of London*, 3 vols (Oxford: Clarendon Press, 1964–72), vol. 1, p. 52.
2. Clark and Cook, vol. 1, pp. 152, 157, 148.
3. John Cotta, *A Short Discouerie of the Vnobserued Dangers of Seuerall Sorts of Ignorant and Vnconsiderate Practisers of Physicke* (London, 1612), p. 10.
4. Mary Lindemann, *Medicine and Society in Early Modern Europe*, 2nd edn (Cambridge: Cambridge University Press, 2010), p. 121.
5. Mary Floyd-Wilson, *Occult Knowledge, Science, and Gender on the Shakespearean Stage* (Cambridge: Cambridge University Press, 2013), p. 37.
6. Juan Huarte, *Examen de Ingenios: The examination of mens wits*, trans. Richard Carew (London, 1594) p. 298.
7. Gail Kern Paster, *The Body Embarrassed: Drama and the Disciplines of Shame in Early Modern England* (Ithaca: Cornell University Press, 1993) and *Humoring the Body: Emotions and the Shakespearean Stage* (Chicago and London: University of Chicago Press, 2004); Michael C. Schoenfeldt, *Bodies and Selves in Early Modern England: Physiology and Inwardness in Spenser, Shakespeare, Herbert, and Milton* (Cambridge: Cambridge University Press, 1999).
8. Gail Kern Paster, Katherine Rowe, and Mary Floyd-Wilson, 'Introduction: Reading the Early Modern Passions', in *Reading the Early Modern Passions: Essays in the Cultural History of Emotion*, ed. Gail Kern Paster, Katherine Rowe and Mary Floyd-Wilson (Philadelphia: University of Pennsylvania Press, 2004), pp. 1–20, p. 15. Two other examples are Katharine A. Craik and Tanya Pollard eds, *Shakespearean Sensations: Experiencing Literature in Early Modern England* (Cambridge: Cambridge University Press, 2013) and Richard Meek and Erin Sullivan eds, *The Renaissance of Emotion: Understanding Affect in Shakespeare and His Contemporaries* (Manchester: Manchester University Press, 2015).
9. See Stephen Greenblatt, *Shakespearean Negotiations: The Circulation of Social Energy in Renaissance England* (Berkeley: University of California Press, 1988), pp. 66–93; Thomas Laqueur, *Making Sex: Body and Gender from the Greeks to Freud* (Cambridge: Harvard University Press, 1992); Janet Adelman, 'Making Defect Perfection: Shakespeare and the One-Sex Model', in *Enacting Gender on the English Renaissance Stage*, ed. Viviana Comensoli and Anne Russell (Urbana: University of Illinois Press, 1999), pp. 23–52. More recent examples are Eve Keller, *Generating Bodies and Gendered Selves: The Rhetoric of Reproduction in Early Modern England* (Seattle: University of Washington Press, 2007) and Kaara L.

Peterson, *Popular Medicine, Hysterical Disease, and Social Controversy in Shakespeare's England* (London: Routledge, 2010).

10. Kim F. Hall, *Things of Darkness: Economies of Race and Gender in Early Modern England* (Ithaca: Cornell University Press, 1995); Ania Loomba, *Shakespeare, Race, and Colonialism* (Oxford: Oxford University Press, 2002).

11. See the special issues on 'Disabled Shakespeares' in *Disability Studies Quarterly* 29.4 (2009), ed. Allison P. Hobgood and David Houston Wood, and on 'Early Modern Trans Studies' in *Journal for Early Modern Cultural Studies* 19.4 (2019).

12. Valerie Traub ed., *The Oxford Handbook of Shakespeare and Embodiment: Gender, Sexuality, and Race* (Oxford: Oxford University Press, 2016).

13. Lindemann, p. 13.

14. The implications of Galenic medicine for an understanding of Shakespeare are comprehensively set out by F. David Hoeniger in *Medicine and Shakespeare in the English Renaissance* (Newark: University of Delaware Press; London: Associated University Presses, 1992), to which the current chapter is much indebted.

15. Ofer Gal, *The Origins of Modern Science: From Antiquity to the Scientific Revolution* (Cambridge: Cambridge University Press, 2021), p. 273.

16. Galen, *On Temperaments; On Non-uniform Distemperment; The Soul's Traits Depend on Bodily Temperament*, trans. Ian Johnston (Cambridge, Mass: Harvard University Press, 2020), p. 347.

17. René Descartes, *Meditations*, trans. John Veitch, in *The Rationalists* (Garden City, New York: Dolphin, 1960), p. 172.

18. Hoeniger, pp. 88, 92, 78.

19. Helkiah Crooke, *Mikrokosmographia: A Description of the Body of Man* (London, 1615), p. 174; see also Paster, *Body Embarrassed*, p. 72.

20. Erin Sullivan, *Beyond Melancholy: Sadness and Selfhood in Renaissance England* (Cambridge: Cambridge University Press, 2016), p. 21.

21. Galen, *On the Usefulness of the Parts of the Body*, trans. Margaret Tallmadge May, 2 vols (Ithaca: Cornell University Press, 1968), 4.13, quoted in Hoeniger, p. 141.

22. Pierre de La Primaudaye, *The Second Part of the French Academie*, trans. Thomas Bowes (London, 1594), title page and pp. 146, 161.

23. La Primaudaye, pp. 215, 229.

24. Catherine Belling, 'Infectious Rape, Therapeutic Revenge: Bloodletting and the Health of Rome's Body', in *Disease, Diagnosis, and Cure on the Early Modern Stage* ed. Stephanie Moss and Kaara L. Peterson (Aldershot: Ashgate, 2004), pp. 113–132, p. 117.

25. La Primaudaye, p. 231.

26. La Primaudaye, pp. 137–138.

27. Gal, pp. 274–276.

28. Nancy G. Siraisi, *Medieval and Early Renaissance Medicine: An Introduction to Knowledge and Practice* (Chicago: University of Chicago Press, 1990), p. 8.

29. Andrew Boorde, *The Breuiary of Helthe* (London, 1547), fols 4^v-5^r.

30. For a discussion of this, see William Shakespeare, *Hamlet*, ed. Harold Jenkins (London: Methuen, 1982), pp. 106–108.

31. Robert Copland (attr.), *The Shepardes Kalender* (London, 1570?), sigs H4^r-I5^v (H6^r).

32. Thomas Elyot, *The Castel of Helthe* (London, 1539), fol. 1^r; Siraisi, p. 101.

33. Elyot, fols 53^v, 64^v.

34. Elyot, fol. 2^r.

35. Elyot, fols 17^{r-v}.

36. Hoeniger, p. 164.

37. Levinus Lemnius, *The Touchstone of Complexions*, trans. Thomas Newton (London, 1576), fol. 133^r.

38. Elyot, fol. 73^v.

39. References are to *The Cambridge Edition of the Works of Ben Jonson*, gen. eds David Bevington, Martin Butler, and Ian Donaldson, 7 vols (Cambridge: Cambridge University Press, 2012), vol. 1.

40. La Primaudaye, p. 18.

41. E. M. W. Tillyard, *Shakespeare's History Plays* (London: Chatto and Windus, 1944), p. 265; Greenblatt, *Shakespearean Negotiations*, p. 49.

42. Schoenfeldt, p. 11.

43. William Vaughan, *Naturall and Artificial Directions for Health* (London, 1600), sigs. C2^r, C7^r.

44. David Hillman, *Shakespeare's Entrails: Belief, Scepticism and the Interior of the Body* (Houndmills: Palgrave Macmillan, 2007), p. 49.

45. Elyot, fols. 22^v, 26^v, 60^v.

46. Paster, *Humoring the Body*, p. 113.

47. Paster, *Humoring the Body*, p. 66.

48. Margaux Deroux, 'The Blackness Within: Early Modern Color-Concept, Physiology and Aaron the Moor in Shakespeare's *Titus Andronicus*', *Mediterranean Studies*, 19 (2010): 86–101 (89).

49. Loomba, *Shakespeare, Race, and Colonialism*, p. 54. See also Robert Burton, *The Anatomy of Melancholy*, ed. Angus Gowland (London: Penguin, 2021), Partition 3, Section 3, Member 1, Subsection 2 (p. 931).

50. Mary Floyd-Wilson, *English Ethnicity and Race in Early Modern Drama* (Cambridge: Cambridge University Press, 2003), p. 33, citing *An Appraisal of the Geographical Works of Albertus Magnus and His*

Contributions to Geographical Thought, ed. Jean Paul Tilman, Michigan Geographical Publication 4 (1971): 101-2.

51. Floyd-Wilson, *English Ethnicity*, pp. 2, 31.

52. Hall, p. 94.

53. Huarte, p. 199.

54. George Best, *A True Discourse of the Late Voyages of Discouerie* (London, 1578), p. 29.

55. See Scott Oldenburg, 'The Riddle of Blackness in England's National Family Romance', *Journal for Early Modern Cultural Studies* 1.1, Spring/Summer 2001: 46–62 (50)

56. Ania Loomba, 'Identities and Bodies in Early Modern Studies', in Traub ed., 228–246, p. 242.

57. Compare *The Taming of the Shrew*, 4.6.46–48: 'Pardon, old father, my mistaking eyes | That have been so bedazzled with the sun | That everything I look on seemeth green.'

58. See also Howard Marchitello, 'Vesalius' *Fabrica* and Shakespeare's *Othello*: Anatomy, Gender and the Narrative Production of Meaning', *Criticism* 35 (1993): 529–558, which links Othello's stated desire to dismember Desdemona to anatomization.

59. Jonathan Sawday, *The Body Emblazoned: Dissection and the Human Body in Renaissance Culture* (Abingdon: Routledge, 1995), p. 56.

60. Siraisi, pp. 86–89.

61. Vivian Nutton, 'Hellenistic and Roman Medicine', in *The Cambridge History of Science*, vol. 1: *Ancient Science*, ed. Alexander Jones and Liba Taub (Cambridge: Cambridge University Press, 2018), pp. 316–344, pp. 325, 342.

62. Siraisi, pp. 89–96.

63. Gal, p. 302.

64. Andreas Vesalius, *De humani corporis fabrica* (Basle, 1543), sigs *4^r ('consectiones propriis manibus obeundas'), *3^v ('dissectionis professorum facile principiis'; 'multo saepius quam ducenties a uera partium humanae harmoniae, usus functionisque descriprione, Galenum declinasse'; 'dissectum corpus operum Naturae studiosis ob oculos collocent'). Author's translation. On the rete mirabile, see p. 621.

65. Sawday, pp. 42, 71, 68.

66. Sawday, pp. ix, 75; Hillman, p. 43.

67. For *Alcazar*, see *The Stukeley Plays*, ed. Charles Edelman (Manchester: Manchester University Press, 2005), p. 96n.

68. Edward More, *The Defence of Women* (London, 1560), sig. A2^v.

69. Hillman, pp. 104, 82.

70. Matthew Steggle, *Laughing and Weeping in Early Modern Theatres* (Aldershot: Ashgate, 2007), p. 125.

71. Thomas Browne, *Religio Medici* (London, 1642), p. 127; William Prynne, *Hystrio-mastix. The Player's Scourge, or Actors Tragedie* (London, 1633), p. 299. Quoted Steggle pp. 89, 65.

72. Alison P. Hobgood, *Passionate Playgoing in Early Modern England* (Cambridge: Cambridge University Press, 2014), p. 10.

73. Glynne Wickham, Herbert Berry, and William Ingram, eds, *English Professional Theatre, 1530–1660* (Cambridge: Cambridge University Press, 2000), p. 81.

74. Thomas White, *A Sermo[n] Preached at Pawles Crosse on Sunday the Thirde of Nouember 1577* (London, 1578), p. 47.

75. Elyot, fol. 8^r.

76. Jonathan Gil Harris, "'Some love that drew him oft from home": Syphilis and International Commerce' in *The Comedy of Errors*, in Moss and Peterson eds, pp. 69–92, p. 81.

77. Cook, p. 422.

78. Girolamo Fracastoro, *Syphilis*, 1.119–124, in *Latin Poetry*, trans. James Gardner (Cambridge, Mass: Harvard University Press, 2013).

CHAPTER 4

1. Some critics, such as Eugene Ostashevsky, take the fact that 'cipher' tends to refer specifically to zero as indicating that 'crookèd figure' should similarly be read as referring to a zero, not a 1. This is plausible, although strictly speaking it is the 1, not the zero, that denotes a million, while 'cipher' could mean any of the Hindu-Arabic numerals (see *OED* 'cipher', *n.*, 3). Eugene Ostashevsky, 'Crooked Figures: Zero and Hindu-Arabic Notation in Shakespeare's *Henry V*', in *Arts of Calculation: Quantifying Thought in Early Modern Europe*, ed. David Glimp and Michelle R. Warren (New York: Palgrave Macmillan, 2004), pp. 205–228, p. 212.

2. Ostashevsky, p. 208.

3. David McPherson, *Ben Jonson's Library and Marginalia: An Annotated Catalogue*, *Studies in Philology* 71.5, Texts and Studies (1974), p. 43. Jonson refers accurately to Euclid in a note on Aelian's *Tactica*: see p. 23. Quotation from *The Magnetic Lady* refers to *The Cambridge Edition of the Works of Ben Jonson*, gen. eds David Bevington, Martin Butler, and Ian Donaldson, 7 vols (Cambridge: Cambridge University Press, 2012), vol. 6.

4. Mary Thomas Crane, 'How Much Do We Need to Know?', *Shakespeare Studies* 49 (2021): 28–37 (35).

5. Paula Blank, *Shakespeare and the Mismeasure of Renaissance Man* (Ithaca: Cornell University Press, 2006), p. 2.

6. Keith Thomas, 'Numeracy in Early Modern England: The Prothero Lecture', *Transactions of the Royal Historical Society* 37 (1987): 103–132, 105.

7. John Brinsley, *Ludus Literarius: or, the Grammar Schoole* (London, 1612), p. 25.

8. Thomas, 111, 109 (citing Bodleian Lib., MS Aubrey 10, f. 29), 114; Edmund Wingate, *Arithmetique Made Easie* (London, 1630), sigs π4v-A1r.

9. Thomas, 103.

10. See for example John Dee's Preface to Euclid, *The Elements of Geometrie*, trans. Henry Billingsley (London, 1570), sig. d4^{v}. See also Jim Bennett, 'The Mechanical Arts', in *The Cambridge History of Science*, vol. 3: *Early Modern Science*, ed. Katharine Park and Lorraine Daston (Cambridge: Cambridge University Press, 2006), pp. 673–695.

11. Adam Max Cohen, *Shakespeare and Technology: Dramatizing Early Modern Technological Revolutions* (New York: Palgrave Macmillan, 2006), pp. 122, 3.

12. Joseph Jarrett, *Mathematics and the Late Elizabethan Drama* (London: Palgrave Macmillan, 2019), pp. 5–6.

13. Henry S. Turner, *The English Renaissance Stage: Geometry, Poetics, and the Practical Spatial Arts, 1580–1630* (Oxford: Oxford University Press, 2006), pp. 25, 29, 230.

14. Gary Hatfield, 'Metaphysics and the New Science', in *Reappraisals of the Scientific Revolution*, ed. David C. Lindberg and Robert S. Westman (Cambridge: Cambridge University Press, 1990), pp. 93–166, p. 93.

15. *The Controversy on the Comets of 1618*, trans. Stillman Drake and C. D. O'Malley (Philadelphia: University of Pennsylvania Press, 1960), pp. 183–184, qu. in Howard Marchitello and Evelyn Tribble, 'Introduction', in *The Palgrave Handbook of Early Modern Literature and Science*, ed. Howard Marchitello and Evelyn Tribble (London: Palgrave Macmillan, 2017), pp. xxv–xlvi, p. xxix.

16. Blank, p. 2.

17. Thomas Sprat, *The History of the Royal-Society of London, for the Improving of Natural Knowledge* (London, 1667), pp. 4, 62, 113. See also Liza Blake, 'The Grounds of Literature and Science: Margaret Cavendish's Creature Manifesto', in Marchitello and Tribble eds, pp. 3–26.

18. Mary Poovey, *A History of the Modern Fact: Problems of Knowledge in the Sciences of Wealth and Society* (Chicago: University of Chicago Press, 1998), pp. 5–6.

19. Jacob Klein, *Greek Mathematical Thought and the Origin of Algebra*, trans. Eva Brann (New York: Dover, 1968), p. 10.

20. Michele Jaffe, *The Story of O: Prostitutes and Other Good-for-Nothings in the Renaissance* (Cambridge, Mass.: Harvard University Press, 1999), p. 27; Brian Rotman, *Signifying Nothing: The Semiotics of Zero* (New York: St Martin's Press, 1987), p. 61.

21. Charles Seife, *Zero: The Biography of a Dangerous Idea* (London: Souvenir, 2000), p. 19.

22. Robert Kaplan, *The Nothing that Is: A Natural History of Zero* (London: Allen Lane, 1999), p. 11.

23. Jaffe, p. 31.

24. Rotman, p. 9.

25. Seife, p. 67.

26. See Paul Benoit and Francoise Micheau, 'The Arab Intermediary', in *A History of Scientific Thought: Elements of a History of Science*, ed. Michel Serres, translator not specified (Paris: Bordas, 1989; Oxford: Blackwell, 1995), pp. 191–221.

27. Fifteenth-century English translations of Sacrobosco and Villedieu can be found in *The Earliest Arithmetics in English*, ed. Robert Steele, Early English Text Society (London: Oxford University Press, 1922).

28. Seife, pp. 19, 39; Kaplan, p. 96.

29. Steele ed. pp. 3–5.

30. On double-entry book-keeping, see Poovey, esp. pp. 33–65.

31. Linda Woodbridge ed., *Money and the Age of Shakespeare: Essays in New Economic Criticism* (New York: Palgrave Macmillan, 2003); Peter Grav, *Shakespeare and the Economic Imperative: 'What's aught but as'tis valued?'* (New York: Routledge, 2008); Natasha Korda, 'Dame Usury: Gender, Credit, and (Ac) counting in the Sonnets and *The Merchant of Venice*', *Shakespeare Quarterly* 60 (2009): 129–153; 130.

32. Korda, 143.

33. For an account of Baskerville's career, see Eva Griffith, *A Jacobean Company and its Playhouse: The Queen's Servants at the Red Bull Theatre (*c. *1605–1619)* (Cambridge: Cambridge University Press, 2013).

34. Brinsley, p. 26. The online English Short Title Catalogue at http://estc.bl.uk lists 23 printings from 1543 to 1610 and 23 from 1615 to 1699.

35. The Golden Rule is used to answer questions such as, 'if 6 articles cost 10 pence what would 9 articles cost at the same price?'. See John Denniss and Fenny Smith, 'Robert Recorde and His Remarkable Arithmetic', in *Robert Recorde: The Life and Times of a Tudor Mathematician*, ed. Gareth Roberts and Fenny Smith (Cardiff: University of Wales Press, 2012), pp. 25–38, p. 30; on Part 3, see p. 31.

36. Robert Recorde, *The Ground of Artes* (London, 1543), fols 2^v, 9^{r-v}.

37. Rotman, p. 78. Eugene Ostashevsky is one critic who reads *Henry V* in relation to Recorde; see 'Crooked Figures', pp. 209–210.

38. Glynne Wickham, Herbert Berry, and William Ingram, eds, *English Professional Theatre, 1530–1660* (Cambridge: Cambridge University Press, 2000), p. 494.

39. Euclid, *The Elements of Geometrie*, trans. Henry Billingsley (London, 1570), fol. 3^r.

40. Jacqueline Stedall, '*The Pathway to Knowledg* and the English Euclidean Tradition', in Roberts and Smith eds, pp. 57–72, p. 57.

41. Robert Recorde, *The Pathvvay to Knowledg* (London, 1551), sig. †2r.

42. Patricia Parker, 'Cassio, Cash, and the "Infidel 0": Arithmetic, Double-entry Bookkeeping, and Othello's Unfaithful Accounts', in *A Companion to the Global Renaissance: English Literature and Cultuure in the Era of Expansion*, ed. Jyotsna Singh (Malden: Blackwell, 2013), pp. 223–241, p. 226.

43. Euclid, fol. 1^{r-v}.

44. Recorde, *The Pathvvay to Knowledg*, sig. A3^v.

45. Jaffe, p. 51.

46. Recorde, *The Ground of Artes*, fol. 61^r.

47. Recorde, *The Ground of Artes*, fol. 64^{r-v}.

48. Shankar Raman, 'Death by Numbers: Counting and Accounting in *The Winter's Tale*', in *Alternative Shakespeares 3*, ed. Diana E. Henderson (London: Routledge, 2008), pp. 158–180, p. 175.

49. Blank, p. 128.

50. Thomas Overbury, *New and Choise Characters, of Seuerall Authors Together with that Exquisite and Unmatcht Poeme, The Wife* (London, 1615), sigs. M5^v-M6^r.

General

An essential volume for anyone researching early modern science is Katharine Park and Lorraine Daston eds, *The Cambridge History of Science*, vol. 3: *Early Modern Science* (Cambridge: Cambridge University Press, 2006), which includes chapters by leading scholars on central topics in the subject. A longer chronological survey is offered in Michel Serres ed., *A History of Scientific Thought: Elements of a History of Science*, translator not specified (Paris: Bordas, 1989; Oxford: Blackwell, 1995). Some excellent shorter surveys are Peter Dear, *Revolutionizing the Sciences: European Knowledge and its Ambitions, 1500–1700* (Houndmills: Palgrave, 2001) and Ofer Gal, *The Origins of Modern Science: From Antiquity to the Scientific Revolution* (Cambridge: Cambridge University Press, 2021). Herbert Butterfield, *The Origins of Modern Science*, rev. ed. (London: G. Bell and Sons, 1957) is a good example of a more old-fashioned approach to the subject; for challenges to the attitude he typifies, see David C. Lindberg and Robert S. Westman eds, *Reappraisals of the Scientific Revolution* (Cambridge: Cambridge University Press, 1990) and Steven Shapin, *The Scientific Revolution* (Chicago: University of Chicago Press, 1996). A lively and accessible account of medieval science is provided in Seb Falk, *The Light Ages: A Medieval Journey of Discovery* (London: Allen Lane, 2020). The ongoing influence of Aristotle is a recurrent theme of the current book; on this, see Charles B. Schmitt, *Aristotle and the Renaissance* (Cambridge, MA: Harvard University Press, 1983).

On the more specific topic of early modern literature and science, Howard Marchitello and Evelyn Tribble eds, *The Palgrave Handbook of Early Modern Literature and Science* (London: Palgrave Macmillan, 2017) is an invaluable collection of essays, as are Juliet Cummins and David Burchell eds, *Science, Literature and Rhetoric in Early Modern England* (Aldershot: Ashgate, 2007) and the special issues of *South Central Review* 26 (2009) on Shakespeare and Science edited by Carla Mazzio and of *Shakespeare Studies* 49 (2021) on Literature and Science edited by Shankar Raman. Howard Marchitello, *The Machine in the Text: Science and Literature in the Age of Shakespeare and Galileo* (Oxford: Oxford University Press, 2011) and Mary Thomas Crane, *Losing Touch with Nature: Literature and the New Science in Sixteenth-Century England* (Baltimore: Johns Hopkins University Press, 2014) are two wide-ranging monographs that include substantial discussion of Shakespeare, among others. Katherine Walker, *Shakespeare and Science: A Dictionary* (London: Arden Shakespeare, 2022) is a reference work with alphabetized entries on Shakespeare's use of scientific terms and concepts. On science in Shakespeare's London, see Adam Max Cohen, *Shakespeare and Technology: Dramatizing Early Modern Technological Revolutions* (New York: Palgrave Macmillan, 2006) and Deborah E. Harkness, *The Jewel House: Elizabethan London and the Scientific Revolution* (New Haven: Yale University Press, 2007).

All of the above contain discussion of the more specific topics detailed below.

Things in Heaven and Earth

Two old but useful handbooks are Francis R. Johnson, *Astronomical Thought in Renaissance England: A Study of the English Scientific Writings from 1500 to 1645* (Baltimore: Johns Hopkins Press, 1937) and S. K. Heninger, *A Handbook of Renaissance Meteorology: With Particular Reference to Elizabethan and Jacobean Literature* (Durham, N. C.: Duke University Press, 1960). For more recent accounts, see Edward Grant, *Planets, Stars, and Orbs: The Medieval Cosmos, 1200–1687* (Cambridge: Cambridge University Press, 1996) and Craig Martin, *Renaissance Meteorology: Pomponazzi to Descartes* (Baltimore: Johns Hopkins University Press, 2011).

On Shakespeare and weather, see Gwilym Jones, *Shakespeare's Storms* (Manchester: Manchester University Press, 2015) and Sophie Chiari, *Shakespeare's Representation of Weather, Climate and Environment: The Early Modern 'Fated Sky'* (Edinburgh: Edinburgh University Press, 2019). Two ecocritical texts that include discussion of weather are Gabriel Egan, *Green Shakespeare: From Ecopolitics to Ecocriticism* (London: Routledge, 2006) and Todd A. Borlik, *Ecocriticism and Early Modern English Literature: Green Pastures* (New York: Routledge, 2011). On the Shakespearean cosmos, see Kristen Poole, *Supernatural Environments in Shakespeare's England: Spaces of Demonism, Divinity and Drama* (Cambridge: Cambridge University Press, 2011).

Matter

For a discussion of the influence of atomist ideas on early modern science, see Robert Hugh Kargon, *Atomism in England from Hariot to Newton* (Oxford: Clarendon Press, 1966). Gerard Passannante, *The Lucretian Renaissance: Philology and the Afterlife of Tradition* (Chicago: University of Chicago Press, 2011) is more literary in its focus; see also the same author's *Catastrophizing: Materialism and the Making of Disaster* (Chicago: University of Chicago Press, 2010). On the influence of Paracelsus, see Charles Webster, *From Paracelsus to Newton: Magic and the Making of Modern Science* (Cambridge: Cambridge University Press, 1982).

Shakespeare's use of Ovid is a vast topic; two learned, accessible introductions are Jonathan Bate, *Shakespeare and Ovid* (Oxford: Clarendon Press, 1993) and Colin Burrow, *Shakespeare and Classical Antiquity* (Oxford: Oxford University Press, 2013). Margaret Healy, *Shakespeare, Alchemy and the Creative Imagination: The Sonnets and 'A Lover's Complaint'* (Cambridge: Cambridge University Press, 2011) argues vigorously for Shakespeare's familiarity with alchemical terms; a contrasting approach is that of Katherine Eggert in *Disknowledge: Literature, Alchemy, and the End of Humanism in Renaissance England* (Philadelphia: University of Pennsylvania Press, 2015), which argues that alchemy in early modern England connoted obsolete but attractive knowledge. On Shakespeare and Lucretius, see the work of Jonathan Pollock, e.g. 'Shakespeare and the Atomist Heritage', in *The Circulation of Knowledge in Early Modern English Literature*, ed. Sophie Chiari (Farnham: Ashgate, 2015), pp. 47–56; for a recent account, see Sean Ferrier and Lisa Walters, 'Hamlet and Lucretian Anxiety', *Shakespeare* 18 (2022): 176–196. On Shakespeare, science, and American colonization, see Denise Albanese, *New Science, New World* (Durham, N. C.: Duke University Press, 1996) and B. J. Sokol, *A Brave New World of Knowledge: Shakespeare's* The Tempest *and Early Modern Epistemology* (Madison: Fairleigh Dickinson University Press; London: Associated University Presses, 2003).

The Body

Two invaluable historical works are Nancy G. Siraisi's *Medieval and Early Renaissance Medicine: An Introduction to Knowledge and Practice* (Chicago: University of Chicago Press, 1990) and the more sociologically focused *Medicine and Society in Early Modern Europe*, 2nd edn (Cambridge: Cambridge University Press, 2010), by Mary Lindemann. A comprehensive summary of the medical beliefs and practices that informed Shakespeare's work is provided by David Hoeniger in *Medicine and Shakespeare in the English Renaissance* (Newark: University of Delaware Press; London: Associated University Presses, 1992).

An idea of the wide range of approaches critics have taken to Shakespeare and the body over recent decades can be gained from Valerie Traub ed., *The Oxford Handbook of Shakespeare and Embodiment: Gender, Sexuality, and Race* (Oxford: Oxford University Press, 2016). Important collections on this topic include David Hillman and Carla Mazzio eds, *The Body in Parts: Fantasies of Corporeality in Early Modern Europe* (New York, Routledge, 1997) and Stephanie Moss and Kaara L. Peterson eds, *Disease, Diagnosis, and Cure on the Early Modern Stage* (Aldershot: Ashgate, 2004). On the humoral body specifically, foundational works for Shakespeare studies are Gail Kern Paster's *The Body Embarrassed: Drama and the Disciplines of Shame in Early Modern England* (Ithaca: Cornell University Press, 1993) and *Humoring the Body: Emotions and the Shakespearean Stage* (Chicago and London: University of Chicago Press, 2004), and Michael C. Schoenfeldt's *Bodies and Selves in Early Modern England: Physiology and Inwardness in Spenser, Shakespeare, Herbert, and Milton* (Cambridge: Cambridge University Press, 1999). On the passions, see Gail Kern Paster, Katherine Rowe, and Mary Floyd-Wilson eds, *Reading the Early Modern Passions: Essays in the Cultural History of Emotion* (Philadelphia: University of Pennsylvania Press, 2004); Richard Meek and Erin Sullivan eds, *The Renaissance of Emotion: Understanding Affect in Shakespeare and His Contemporaries* (Manchester: Manchester University Press, 2015); and Erin Sullivan, *Beyond Melancholy: Sadness and Selfhood in Renaissance England* (Cambridge: Cambridge University Press, 2016). On the passions and literary response, see Matthew Steggle, *Laughing and Weeping in Early Modern Theatres* (Aldershot: Ashgate, 2007); Katharine A. Craik and Tanya Pollard eds, *Shakespearean Sensations: Experiencing Literature in Early Modern England* (Cambridge: Cambridge University Press, 2013); and Alison P. Hobgood, *Passionate Playgoing in Early Modern England* (Cambridge: Cambridge University Press, 2014). Suparna Roychoudhury's *Phantasmatic Shakespeare: Imagination in the Age of Early Modern Science* (Ithaca: Cornell UP, 2018) includes discussion of early modern notions of the brain and its workings.

Accounts of how racial difference was understood in early modern England can be found in Kim F. Hall, *Things of Darkness: Economies of Race and Gender in Early Modern England* (Ithaca: Cornell University Press, 1995), Ania Loomba, *Shakespeare, Race, and Colonialism* (Oxford: Oxford University Press, 2002), and Mary Floyd-Wilson, *English Ethnicity and Race in Early Modern Drama* (Cambridge: Cambridge University Press, 2003). On anatomy and dissection, see Jonathan Sawday, *The Body Emblazoned: Dissection and the Human Body in Renaissance Culture* (Abingdon: Routledge, 1995) and David Hillman, *Shakespeare's Entrails: Belief, Scepticism and the Interior of the Body* (Houndmills: Palgrave Macmillan, 2007). Finally, on the practice of medicine see Todd H. J. Pettigrew, *Shakespeare and the Practice of Physic: Medical Narratives on the Early Modern English Stage* (Newark: University of Delaware Press, 2007) and Kaara L. Peterson, *Popular Medicine, Hysterical Disease, and Social Controversy in Shakespeare's England* (London: Routledge, 2010); Mary Floyd-Wilson's *Occult Knowledge, Science, and Gender on the Shakespearean Stage* (Cambridge: Cambridge University Press, 2013) focuses a different way of acting on the body, namely that facilitated by the hidden properties of objects and persons.

Maths

Keith Thomas offers an excellent survey of 'Numeracy in Early Modern England: The Prothero Lecture', *Transactions of the Royal Historical Society*, 37 (1987): 103–132. Accessible historical introductions to zero and its significance are provided in Robert Kaplan, *The Nothing that Is: A Natural History of Zero* (London: Allen Lane, 1999) and Charles Seife, *Zero: The Biography of a Dangerous Idea* (London: Souvenir, 2000). More literary and theoretical approaches are taken in Brian Rotman, *Signifying Nothing: The Semiotics of Zero* (New York: St Martin's Press, 1987) and Michele Jaffe, *The Story of O: Prostitutes and Other Good-for-Nothings in the Renaissance* (Cambridge, Mass.: Harvard University Press, 1999). Mary Poovey, *A History of the Modern Fact: Problems of Knowledge in the Sciences of Wealth and Society* (Chicago: University of Chicago Press, 1998) considers the growing authority ascribed to numbers; the essays in David Glimp and Michelle R. Warren eds, *Arts of Calculation: Quantifying Thought in Early Modern Europe* (New York: Palgrave Macmillan, 2004) similarly focus on the impulse to quantify and include Eugene Ostashevsky's chapter on *Henry V*. Gareth Roberts and Fenny Smith's collection *Robert Recorde: The Life and Times of a Tudor Mathematician* (Cardiff: University of Wales Press, 2012) provides a good introduction to a figure who appears repeatedly throughout the present book.

Turning to the drama specifically, Henry S. Turner, *The English Renaissance Stage: Geometry, Poetics, and the Practical Spatial Arts, 1580–1630* (Oxford: Oxford University Press, 2006) is a rich and substantial account of the two-way relationship between the theatre and applied mathematics. Paula Blank, *Shakespeare and the Mismeasure of Renaissance Man* (Ithaca: Cornell University Press, 2006) addresses Shakespeare's ambivalent relationship to quantification. For introductions to the New Economic Criticism, see Linda Woodbridge ed., *Money and the Age of Shakespeare: Essays in New Economic Criticism* (New York: Palgrave Macmillan, 2003) and Peter Grav, *Shakespeare and the Economic Imperative: "What's aught but as'tis valued?"* (New York: Routledge, 2008). For discussion of mathematical and accounting terms as they figure in specific texts, see Shankar Raman, 'Death by Numbers: Counting and Accounting in *The Winter's Tale*', in Diana E. Henderson ed., *Alternative Shakespeares 3* (London: Routledge, 2008), pp. 158–180; Natasha Korda, 'Dame Usury: Gender, Credit, and (Ac) counting in the Sonnets and *The*

Merchant of Venice', *Shakespeare Quarterly* 60 (2009): 129–153; and Patricia Parker, 'Cassio, Cash, and the "Infidel o": Arithmetic, Double-entry Bookkeeping, and *Othello*'s Unfaithful Accounts' in Jyotsna Singh ed., *A Companion to the Global Renaissance: English Literature and Culture in the Era of Expansion* (Malden: Blackwell, 2013), pp. 223–241. Finally, a recent monograph devoted entirely to the topic is Joseph Jarrett, *Mathematics and the Late Elizabethan Drama* (London: Palgrave Macmillan, 2019).